HOW TO CHANGE A MEMORY

How to Change a Memory

ONE NEUROSCIENTIST'S
QUEST TO ALTER THE PAST

STEVE RAMIREZ

PRINCETON UNIVERSITY PRESS

PRINCETON & OXFORD

Published by Princeton University Press
41 William Street, Princeton, New Jersey 08540

press.princeton.edu

GPSR Authorized Representative: Easy Access System Europe — Mustamäe tee 50, 10621 Tallinn, Estonia, gpsr.requests@easproject.com

All Rights Reserved

ISBN 9780691266688
ISBN (e-book) 9780691274294

Editorial: Hallie Stebbins and Chloe Coy
Production Editorial: Ali Parrington
Jacket Design: Heather Hansen
Production: Erin Suydam
Publicity: Matthew Taylor
Copyeditor: Susan Matheson

Jacket Image Credit: (*front*) by Heather Hansen; (*back flap*) SA-10 / Shutterstock.

This book has been composed in Arno

Printed in the United States of America

10 9 8 7 6 5 4 3 2 1

For Xu

CONTENTS

ACKNOWLEDGMENTS

WHEN I was a kid, I didn't always want to be a professor or a scientist. I didn't really know what these even meant. Professors taught things like chemistry and Shakespeare, and scientists wore white lab coats and stared at different-colored liquids all day. Both sounded like the opposite of fun, so instead, I wanted to be the world's greatest Pokémon master. I collected and protected every single card I could get my hands on and guarded each one as if it were my firstborn. If my parents had any extra money after doing our food shopping for the week, we'd go to a card store next to the supermarket, and I'd ask the cashier if I could hold a pack of cards up to my ear so that I could try and *listen* for that rare holographic card that might be communicating to me (it never really worked).

When the card store closed, a local bookstore opened up. But luckily they also sold Pokémon, and because I "gotta catch 'em all" (the Pokémon slogan), I now had a new weekly destination. Little did I know that the books would lure me in as well.

Some books caught my attention just because they sounded cool. What was all the buzz about *The Da Vinci Code*? Did I need to read the Bible first to understand it? Would looking through books of Renaissance art help? Why was everyone mad at *The God Delusion*? How did *A Brief History of Time* sell a gazillion copies? All these questions stemmed from my disbelief that one person could possibly know so much and have so much to say that it required an entire freaking *tome* to get the ideas across.

It could've been that I subconsciously found the challenge tempting, or that somewhere deep inside me there was a whisper of confidence that wanted to turn into a roar of productivity to write a story, *any story,* worthy of being shared. I can describe this feeling today in the same way that I would when I was in elementary school: when leaving a bookstore, I had a stubborn, determined, and unwavering belief that I'd say to myself, *I will belong here one day.*

I have always wanted to write a book. My parents probably figured this out long before me since I'd annoy them at the end of every school week to drive me to the local bookstore so that I could get lost in whatever topic happened to be on my mind—from Pokémon to superheroes in elementary school, to music and history in middle school, to religion and science in high school. When I was old enough to get my driver's license, I'd ask my friends if they wanted to go to the mall—where I'd spend most of our collective time at the bookstore while they begged for us to go literally anywhere else. When my friends said that they were too busy to go with me because they actually had lives, I'd drive to the bookstore on my own and spend entire afternoons there, investigating whatever was of interest in that moment. Being surrounded by so much knowledge meant that I was at the forefront of what humanity knows and doesn't know; I wanted to connect some new dots to add to our understanding of the world and ourselves. This really mattered to me because it felt like the kind of important contribution to society that, well, gets you placed on the bookshelves.

Even throughout college and graduate school, writing a book seemed to be exactly the kind of "impossible" that I knew I wanted to accomplish one day. The only part missing was the exact topic I'd write about (kind of a key ingredient, it turns out!), which didn't hit me until I started to see a much deeper value in my own life experiences, especially as the darker topics of death, anxiety, and addiction began tugging in the other direction and devaluing the very things that were giving me a voice. Writing this book became a way to tug back and to put these life experiences into focus so that I could learn from them one memory at a time. Call it therapeutic, call it getting into grammatical shape, call it storytelling. I just call it taking an essential part of my personhood and putting it in a form that other people can understand. I truly hope this book connects with you, and if it does in any way, then we've connected some of the most spectacular dots that exist. In turn, I hope that we see a bit of each other in one another.

So I thank you, dear reader, from the bottom of my hippocampus, for letting a core part of myself into your life through this book. *How to Change a Memory* is a literal dream come true.

To the Liu family: You brought a son into this world who changed neuroscience and who changed my life. I respect you all so much and hold you in my heart. I promise to lastingly honor Xu and his memory.

To my mom and dad, Delmy Moreno and Pedro Ramirez: *Gracias por todo mami y papi. Los quero mucho.* It's a privilege to be copassengers in life and a blessing to be your son. You've taught me how to breathe through every season of life with hope and love.

To my sister and brother, Jessy Pervez and Marvin Ramirez: Growing up, I never really had "role models" out in the real world who looked like me, especially in science. Imagine my surprise when you both decided to be lawyers! Close enough. At least the three of us happen to be genetically related. Thank you for being my role models throughout life (including stealing Dad's car at night!). You both mean the world to me. To Kazi Pervez and Sheree Ramirez: Thank you for always looking after me from childhood to today and for being an endless source of laughter, resilience, and perspective. I'm so grateful that we're constants in each other's lives.

To Melisa, Daniela, and Sabrina Demaestri: I can honestly say that my soul is nourished and my spirits are uplifted every single time we hang out. Our trips, runs, and game nights together are the *good stuff* in life, and I'm so grateful to have you as sisters-in-law. The three of you add melody and compassion to life, and I look up to you so much. Graziella Chamme and Nathaniel Schub: Thank you for adding vibrance and endless warmth (and, ironically enough, for helping me turn down the AC when no one is looking) to all the times we hang. I love you all and thank you for battling all of life's territorial Tammys together!

To my nieces and nephews, Johan, Julien, Sofia, Frank, and Anthony: Keep doing all of what brings you joy—boxing, drawing, helping others, playing basketball, trying out a new video game. Tio will drive you to the bookstore (or anywhere you want, really!) and get us enough dessert to force us into hibernation. Our family is one giant team, and I'll always be in your corner for all that life has to offer.

To Patricia Sicouly: *Muchisimas gracias* for embracing me as a son-in-law and for sharing your poetic wisdom and love during so many wonderful family trips together. *Un habrazo fuerte.* To Edgardo Demaestri: *Muchisimas gracias* for welcoming me into your family and for showing me how to make sense of all the quantitative and qualitative truths that together build a fulfilling life. I truly enjoy learning from each of you through all the rich and multi-layered life experiences that you've shared.

To my friends and your beloved families, I think of you all daily and look forward to every chance we get to make many more memories together. Birthday parties, barbecues, Texas Roadhouse outings, family vacations, losing our minds at the Sphere, and just texting a bunch of nonsense regularly and pebbling our way through life. Who knows, maybe one day we'll all live in 9R again! Thank you, Erin, Khadija, Maya, and Leah Bashllari; Jasmin, Grace, and Adrian Imsirovic; Walid Elhirach and Ellii Cho; Michael, Cailin, and Mischa Wells; Gabriel Stein and Ariel Azoff; Sarah Leblanc; Travis Rich, Kate Wescott, and Arthur Rich; Maisam Dadgar, Anna Lukes, and Elias Dadgar; Kevin Capelotti and your family. Wah wahh; I love you all!

Thank you to Team X for being my lab family: Briana Chen Anaya, Joanne Zhou, Grace Lin, Pei-Ann Lin Acosta, Amelia Mockett, Elizabeth Shanahan, The UROPean union, Tomás Ryan, Melina Tsitsiklis, Karen Hao, Roger Redondo, Catherine Potter, Autumn Arons, Teryn Mitchell, Michele Pignatelli, Joshua Kim, Takashi Kitamura, Dheeraj Roy, and Arvind Govindarajan. Thank you to Joshua Sariñana, Alex Rivest, Katie Mulroy, Carrie Ragion, Mike Ragion, Jennie Young, Caitlin Vander Weele, Lily Smith, Masha Ledoux, Laura Smith, Emily Hueske Wilson, The Transgenic Dragons, Junghyup Suh, Arek Hamalian, the entire Tonegawa Lab, and the Brain and Cognitive Sciences community at MIT for over half a decade's worth of engrams.

To my late and great mentor, Howard Eichenbaum, thank you for taking a chance on me as an inexperienced undergraduate looking to get some hands-on research experience and for always setting time aside to brainstorm on life, science, and everything in between. To my graduate mentor, Susumu Tonegawa, thank you for modeling a truly fearless and unbounded approach to science, which I try to embody regularly today. To Chris MacDonald: Thank you for being my day-to-day mentor when I was an undergraduate and for being my friend since then.

To the editorial powerhouses who saw the art within the giant slab of untouched marble that this book originally was, once upon a time. Courtney Young, you were my first coach of sorts in writing, and I'm so thankful for all the work we did together for this book. Anyone who gets to work with you is under spectacular editorial guidance, and it was a privilege to learn from someone who is equal parts erudition and compassion.

To Amanda Moon: You believed in me as a writer in exactly the ways I needed to be believed in to keep writing this story. Thank you. I will never forget the "Aha!" moments I experienced while Zooming with you when I was in Italy and suddenly having decades-old memories come flooding back. These memories immediately made their way into the book. You have a magical capacity to unearth parts of a person that are so essential to their being and simultaneously essential to the narrative at hand, and you do so with such TLC. You are one of a kind.

To my inimitable book agent Sarah Levitt, who muscled through the years that I've taken to write this book without ever losing an ounce of faith in me as an author. Sarah, thank you so much for being a constant in my career as a scientist and writer, and for supporting me through every single meeting we've ever had (including the meetings where I had to run out of the room because I was still battling some alcohol demons). I always knew I wanted to write a book, but you helped me find the story, and this was doubly challenging because you managed to find that this story was *within me* all along. I will always thank you for seeing me for who I am. Your knowledge and acceptance of me gave me a newfound sense of strength when writing this book. Brainstorming science and authorship goals with you is such a treat, and you've taught me so much about the art of writing and publishing a book that I owe you my left amygdala if you ever need it. I can't wait for round two!

And a very, very special thank you to Hallie Stebbins, my editor extraordinaire, and Princeton University Press, who truly connected with the vision for this book from day one and brought out my voice with such thoughtfulness and kindheartedness. Hallie, from the very beginning, working with you has been as organic as it has been fulfilling. Thank you for being a pair of ears when I needed guidance, for being a pair of shoulders when I was struggling and stuck with writing (and life), and for being the wisest pair of hands when the writing required it. Every page in this book has benefited from your remarkable ability to juggle both a vision and the written word. You are a real mentor to me in this new adventure of book writing, and I can't wait for our next big project together.

To Dan, my therapist, who for years has taught me that the monster under the bed still exists even if I look away—and that I've had the power in me all along to jab-cross-hook-to-the-body the damn thing. Dan, you helped me navigate my own thoughts and feelings with such compassion rather than with shame and also to understand myself in the context of the relationships around me. Thank you for being a sage-like guide as we untangled the knotted mess that was my brain. You're the man.

To the Recovery Elevator and Café RE community: Thank you for welcoming in a broken version of me with open arms and helping me find a new and healthier path in life. You all taught me to find joy in sobriety, and it's because of every single one of our community members that I truly believe that the opposite of addiction is connection. When I attended my first meeting, and even before I spoke a single word, I already felt at home. It's one of the most inspiring things I've ever felt. You all understood what it meant to live with addiction, and rather than try to problem-solve in the moment or guilt-trip a person, you all made me feel safe, understood, and even protected. We are kindred spirits. I knew after that first meeting that my sobriety journey would only keep moving forward and with the heroic group of people that you all are. I love you all, and I will not drink with you today ♥.

Thank you to the environments that have supported me throughout my career. To Boston University, Michael Hasselmo and the Center for Systems Neuroscience, David Somers, Chantal Stern, the Neurophotonics Center, the Department of Psychological and Brain Science, the

Department of Biomedical Engineering, and to all of my colleagues for supporting me from my beginning as a college student all the way to returning to BU as a professor. Special shoutout to our wagyu Wednesday club—Heidi Meyer, Ben Scott, Mark Howe, Lynne Chantranupong, Laura Lewis, Meg Younger, and Amelia Stanton. Thank you to my cool kid neuroscience colleagues who make the field better through your thoughtful approach to experiments, data, mentorship, and your never-ending quest for delicious dinners: Denise Cai, Tristan Shuman, Christine Denny, Mackenzie Mathis, Maya Opendak, Michael Drew, Gisella Vetere, Bianca Jones Marlin, Christina Kim, and Hector Bravo.

Thank you to MIT for nourishing my inner kid for half a decade and for teaching me the art of not sucking at being a scientist. To the Harvard Society of Fellows: our Monday dinners felt like my adult version of going to the bookstore, and this time I had plenty of friends who wanted to think together about how all our fields are connected. To the National Institutes of Health for literally funding my lab into existence, and to Joshua Sanes at Harvard for donating a behavior room and lab bench where we could get my lab's memory manipulation work officially underway! Thank you deeply to the Ludwig Family Foundation for your continued support and admirably humane approach to science and discovery. To the Pew Scholars Program in the Biomedical Sciences, the Chan-Zuckerberg Initiative, the Air Force Office of Scientific Research, the Brain and Behavior Research Foundation, National Geographic Society, and the McKnight Foundation—thank you all for supporting our work in neuroscience and beyond.

Thank you to the staffs at Trident Bookstore and at Eastern Standard. I wrote the majority of this book while sitting at your bars; and while I wasn't always sitting there looking pretty, your hospitality and kindness lifted my spirits over the years. To our TEDx team: Danielle Duplin, Evan Wondolowski, and the rest of the squad, thank you for helping Xu and me bring our unhinged science ideas to life with your professionalism, creativity, and unwavering dedication.

To the Ramirez Lab, past, present, and future: The best part of being a scientist is the people that make up the field. I admire every one of you for elevating the field with your presence, discoveries, and humanity—we all stand on your shoulders. Thank you, team. Oh, by the way, I still consider myself the best ping pong / foosball / Piano Keys player in all of Boston!

Thank you to all of our mighty mice who make possible the magnificent discoveries in the life sciences. When we have cures for all that ails us, the story of how we got there will always have you at its center, the heroes of biology.

And before I get to the last part, I want to say that, in the same way that I start off this book by mentioning all of what I don't remember about May 26, 2011, because I'm relying on my memory, I'm also relying on my memory here (mostly); so if I forgot anyone, please rest assured there's a dormant engram of you in my brain that's shaping the scientist and person I am. If I forgot you, please yell at me when we meet up next, and bring the engram back to life so that I can re-store it enduringly!

I don't know how else to say this next part: this book has seen some shit—somewhere between loss and addiction and between love and friendship, there is something to be learned. So I want to make today the day when I get to talk to my younger self and say, *Writing a book was even more work than you thought it'd be, but, boy, was it worth it because the kid you are today and the adult you'll be in a few decades will still be you—a happy person who dreams big. Like Mom always says, use your brain but follow your heart. You'll have a lot to say and a lot to share because you have so many feelings, which sometimes don't really make sense; but there's a part of you that believes that if you can make them make sense, then this will somehow be used for good, both for yourself and for anyone who connects with you. Your friends and family and complete strangers will find value in that. Who knows, maybe someone, someday, will randomly go to a bookstore and find your story there, waiting to be rediscovered. You might make them laugh, cry, and resonate with you; they may even disagree with everything you said and use the book as a doorstop! But regardless of what everyone else thinks, just know that I find value in you and I'm proud of you. Oh, and maybe I'm breaking some rules of literary time-travel here, but in your thirties you'll find someone who is even more magnificent than our esteemed Pokémon card collection, and this person will help you "catch 'em all" when it comes to capturing all the moments of happiness that life is about.*

I know it may sound scary, but she will save your life. It is always okay to need some help. There is real life strength in doing so.

That person has seen every single stage of writing this book and more. She is someone that I'm so truly lucky to have met, let alone marry. Thank you to my wife, Camila Demaestri. *Mi amor*, this book is possible because of you, but more importantly, *I am possible* today because of you. I remember leaving work to go together to *work more* at the Lenox, all the way to getting married there, all the way to when we read the introduction together at Delilah's in Vegas. We are each other's homes, and you have my heart entirely. Whenever I look over at the couch and see you there smiling, playing a game, embroidering, watching a show, or doing some data analysis, I stop and remind myself, *This is why we have memories. Te amo.*

A Memory of a Memory

MAY 26, 2011 might have been a sunny day in Boston, perfect for a brisk jog along the Charles River. Or it might have been a gloomy day, just right for a pint and a movie in Harvard Square. It could have been the ideal day for baseball if the Red Sox were back in town. But I can't tell you for sure. I can't tell you what I had for breakfast, lunch, or dinner that day. I can't tell you who I called, what news I read, what music I listened to. There are bits of experience I don't remember because, well, I'm relying on my memory.

And yet, somehow, I do remember that I was in a windowless dark room at the Brain and Cognitive Sciences building at MIT that afternoon, transferring a small, black mouse from the palm of my hand into an almond-scented box about the size of a milk crate, with white floors, transparent walls, and a camera mounted overhead.

The mouse began sniffing its surroundings. It was no stranger to the box, having investigated the same corners just yesterday. The fact that nothing monumental had occurred during its initial journey in the box was important: it meant that the mouse had no reason to be afraid this time around. It could go about its business without fear as I recorded its behavior with my lab partner, Xu Liu.

Ten days earlier, I watched as Xu anesthetized this very same mouse. We both felt similarly regarding the lab mice we worked with: we approached them with veneration for their biological revelations and with tremendous care for the life that they experience. Xu in particular took this relationship seriously, and that day I could tell how much this dynamic meant to him as he carefully lowered two glass barrels into two small holes through the top of the skull of the mouse. Like miniature flashlights, about the width of cocktail straws and shorter than the nail on your pinky finger, these glass barrels are capable of funneling and focusing light onto the part of the mouse's brain in which they are nestled—in this case, in the mouse hippocampus. Why did we want to shine light onto the mouse's hippocampus though? As part of our experiment, we'd made this particular mouse special using some genetic trickery called optogenetics. Put succinctly, optogenetics entails hand-crafting special bits of DNA that make a cell light-sensitive, and delivering those bits into very specific cells into the brain. Once these brain cells are made light sensitive, researchers can turn them on or off with light, much like a switch. Xu and I planned to focus beams of light directly onto our light-sensitive cells in the hippocampus, and voilà, those cells would turn on. It's these particular hippocampus cells, we hypothesized, that contained a memory.

Based on what we knew about the inner workings of the brain, Xu and I had every reason to believe that the hippocampus is like a mental time machine: an area that is active when a mouse is trying to remember the shortest path to return to the tasty crumbs in the kitchen pantry, or in one of us humans, when you recall the memory of your first kiss, or hearing your baby coo for the first time, or last Friday's delicious steak frites dinner. The hippocampus contains millions of brain cells, which chunk space and time into our personally experienced events. The hippocampus, in short, is crucial to the process of memory, in both our mouse and in humans. It teleports us to relive the past.

————

Xu and I were playing with an idea that day in the lab, as our mouse scurried about its box: Could we "turn on" a memory if we triggered the parts of the brain where it lived?

We were in essence testing a hypothesis first put forth over one hundred years ago by the German zoologist Richard Semon. Semon proposed that memories are a kind of lasting physical imprint, or "trace" in the brain—somehow, the marvelous waters of memory carve measurable grooves in the neural riverbeds of the brain. For the savvy enthusiast, there's an official term for this so-called memory trace, and it was first proposed by Semon himself: *engram.*

The holy grail of memory neuroscience, the engram is thought to be the key to unlocking the power of our brain's mental time machine. Once we find the engram, Semon's prescient idea goes, we could probably find a way to reverse engineer memory and, ultimately, control it.

Xu and I used to refer to memory researchers as the auto-mechanics of the brain, taking apart the fleshy machine between our ears one piece at a time in an attempt to understand what each piece does and how it enables our smooth mental time-travel. "If you can break it and make it, then you can understand it," he'd say.

By 2011, memory researchers had already done just that. They had discovered that it was possible to "erase" memories—to "break" them. If scientists could do *that*, well, why couldn't we find memories and reactivate them instead? Breaking something lets us know how it works by preventing some output from happening, Xu and I reasoned, but if we could manually recall a memory—if we could stimulate this same output—then we could get the mental time machine to run again . . . and again and again at will.

Our goal was initially nonscientific in its basis: we'd break into and jump-start the brain's time machine. The project had a name in the lab, one that we felt had a mysterious grandeur to it: Project X. It all sounded pretty sci-fi to us, and that was exactly what hooked us in. The idea of *activating* a memory? Very *Total Recall* with a bit of *Inception* thrown in. We felt like kids again, and our playground was the science lab. Sometimes an idea for an experiment just feels "too important and too damn cool not to do," we'd say.

We thought activating cells that held onto a memory would cause a domino effect in the brain that ultimately led to recollection of that memory. After all, external stimuli in the world do this to us all the time: walking past a bakery and smelling the maple bacon doughnuts might remind you of the last time you cheated on your diet; the sight of a particularly unfashionable shirt at the mall might remind you of an ugly

Christmas sweater party with the family; the smell of tequila might re-mind you of that time you dulled your grief with alcohol and woke up in even more pain the next morning. These sights and sounds and smells force us to relive the worlds of the past. They bring engrams back to life.

We just wanted to bypass the external stimuli.

Say that Xu and I were able to reactivate a memory. What next? One of the central goals of doing this kind of science is to discover funda-mental truths about how the brain works and to use these truths to help people. Our biggest ambitions for our work included applying a new understanding of the mechanics of memory to treat disorders of the brain. We wondered if we could one day suppress a negative mem-ory to prevent the debilitating effects of PTSD, or toggle down a bout of overwhelming anxiety to prevent a panic attack. If we could activate a memory, then we could think of memory as something our brain naturally produces *as well as* a potential antidote that the brain con-tains to rid itself of suffering. The possibilities would be endless: What if we activated positive memories to curb symptoms of depression, or what if we brought a memory back that was thought to be lost to Al-zheimer's, or what if we could etch in entirely new memories to pro-duce a cognitively enhanced brain? All of these possibilities relied on one thing—the ability to control memory, which was what Xu and I had within reach.

Given how far the field had already come, we thought a repair shop for the brain wasn't all that far-fetched.

————

A perceptive, soft-spoken, Shanghai-born scientist, Xu Liu came off as proper, formal—fully buttoning up his lab coat was a rather ritualistic act—and the level of intention and respect he brought to the study of science was like nothing I had ever witnessed. When he was locked into his science, everything else became background noise, as if the only two things in the world that mattered in those moments were him and the experiment at hand. Watching Xu do science felt like watching pure discipline in motion.

But Xu was also a kindhearted mentor, a big brother in the lab. His discipline in the experimental testing rooms transformed into thoughtfulness in his daily mentorship—and it was in this mentorship role that he was *really* in his element. I often felt like Xu had a career-length road map for how he would train me in neuroscience. He told me he decided to take me under his wing and become my day-to-day mentor because our friendship was "organic" from day one in the lab. As he put it: "I knew we'd have chemistry doing science together. Get it?" It was a terrible joke, and yet I couldn't help but laugh.

Xu joined MIT as a postdoc fellow a few years before I arrived there for grad school. He spent years mastering the surgeries and techniques needed to test Semon's engram hypothesis. It was his passion. Late one night, after yet another nail-biting round of Jenga with me, he told me that memory was the most mysterious thing he knew of, but it was a mystery he was confident he could solve with science. In graduate school, he studied how memories are formed and retrieved—and he did this using the tools of molecular biology to study how individual neurons represent memories in the fly brain. Like the human brain, the fly brain also consists of neurons, and it is much easier to play around with fly neurons than human neurons. He moved on to study mouse brains because they're "big fly brains solving mouse problems," as he put it.

But Xu didn't just have the fly, mouse, or even the human brain in mind—no, he was thinking on a much grander scale: Xu wanted to understand the phenomenon of *memory* itself. He believed that memory isn't just recalling what you ate yesterday or your high school graduation; memory is nothing less than the perpetual beating heart of life. It appears everywhere, from single cell organisms to jellyfish, fungi to flies, mice to humans, and from life that started billions of years ago to life that exists today. We are all endowed with this biological machinery capable of preserving what once was. Memory takes many forms, but it is a biological constant.

It follows, then, that knowledge of memory in any organism will inform our grander understanding of how memory works in general and, in some capacity, will be applicable to humans. For Xu and me, rodents were the perfect model organism in which to use our neuroscience tool kit to study how to activate a memory.

We began our experiments by trying to manipulate fear memories. Our reasoning: we know a great deal about the neural circuitry underlying fear in both rodents and humans (it's similar enough), and we know that this circuitry also intersects with a variety of psychiatric disorders, including PTSD and anxiety. So we had a leg up in knowing how fear memories are created and where in the brain to look for them, and we knew that manipulating fear memories could have beneficial effects for patients suffering from fear-based disorders. Fear memories are also incredibly potent; so we figured that when we activated one, we'd know because the mouse's behavior would change immediately, making it a scientifically measurable output. For example, if you think about the impossibly high-pitched noise of a dentist's drill, a hair-raising, "why-god-why" cringe immediately ensues. Your brain made a powerful association—it was *conditioned*—because at some point in the past that noise was followed by pain and pressure while you were told to sit still.

Our first step was to create a fear memory that we could then try to manipulate. One of the most common ways that we can create fear memories in rodents is to place the animals in a new environment—in the lab, this means a small box with a metal grid for a floor, dim lighting, and a black triangular roof scented in this case with a particularly zesty orange aroma—and then send a very mild shock to their feet so they associate this area with an important event. They're thus conditioned: they're trained to associate the environment with a negative stimulus. After they acquired this fear memory, whenever the rodents were placed in the box, they would *freeze* in place to protect themselves from the looming threat of a negative stimulus. Xu and I kept track of how often the mice were freezing as our measurable readout of recalling a fear memory.

Xu and I figured we could activate the brain cells in the hippocampus that produced a fear memory—and that we could do this with light. We used the optogenetic tools Xu had spent so long creating in the lab to finally test Semon's hypothesis. The mouse's brain was primed: its hippocampus cells had been manipulated so that they could be controlled with light, and the tiny glass barrels that would funnel that light had been implanted.

On May 26, 2011, the day after we first created the fear memory, we placed our mice in a "safe" box—a box where nothing had happened, where no shock had been delivered. We began by connecting the glass barrels implanted in the mouse's brain to a laser that emitted a brilliant Avatar-blue light. To turn on the laser, we simply had to click a switch. One *click*—that was all that separated us from a scientific triumph, from a completely puzzling result, or from the tediousness of . . . nothing. But that's science in nutshell: maybe you'll get lucky and see something that has never been seen before; maybe you'll finish the day scratching your head wondering what in the world just happened. Or both.

On the day of our big experiment, we funneled pulses of light through the surgically implanted optic fibers, bombarding the mouse's neural tissue with photons, attempting to awaken the dormant memory. Xu and I sat and watched, eager, nervous, the lab like a command center during the first few seconds of a rocket's lift-off.

We turned on the laser, the mouse's ears perked up—and it immediately stopped moving. It tensed up, vigilant and frozen in place, perhaps experiencing the echoes and murmurs of a fearful engram flicker in and out of its mind.

I finally broke the silence: "So . . . did that . . . just freaking work?"

"We have to run controls," Xu responded briskly. "We have to do the experiment large-scale and double blind. We have to replicate it. But, let's just call it a day and go have a drink across the street because, yes, I think it actually freaking worked."

"I knew it!" I yelled, splashing Xu with the water I was using to clean the floors in the mouse cage. That day, at the end of my first year working with Xu at MIT, I realized that we had made the discovery of a lifetime.

I gave Xu a hug, the first of only two we would share.

———

That Xu and I thought the mouse was recalling a bona fide memory was, of course, anthropomorphic conjecture. No one knows exactly what a memory looks like in the brain, and we certainly don't know what it's like to be a mouse. We do, however, have a rough idea of how cells and mice often behave when they're recalling an experience that was nega-tive. A mouse reliving a negative memory will freeze in place. The objec-tive description of our scientific findings is this: Xu and I turned on the mouse's brain cells that were previously active during the formation of a fear memory, and we caused the mouse to freeze in place. A mouse may never be able to tell us otherwise, but we have very good reason to believe that our objective description can be translated as: we shot pin-point lasers into the mouse's brain that made the mouse relive the fear-provoking memory.

The objective description would also tell you that memory is multi-faceted. It is how the brain processes every experience you've ever had and turns each one into cellular signals that contain information about the past. And what are these "cellular signals" exactly? Well, a wealth of research lets us know that there are processes, stages of memory that transform over time, and we can label them this way: Encoding, Storing, Retrieving, and Updating. All of these processes simply describe the larger phenomenon of memory itself: it is a dynamic integration of the past with the present to permit future livelihood. In other words, mem-ory recalls the past in the present to help us better navigate the future. It is a history with the breadth of all that we've felt and the depth of all that we've experienced. Just as we learn history to learn *from* history, our brain keeps a record of our past so that we can learn from it.

Let's dwell on this definition of memory for a second. For starters, memory isn't *only* about preserving the past. Recalling a memory in the present helps us understand the environment that we find ourselves in, and it helps us decide what to do next. This means that memory has a purpose that is much grander than being a neural photo album of your life. In fact if you gave me five seconds on a talk show to give the world the answer to "What is memory?"—since *everyone* is tuning in to find out—I'd say the following: "Memory is what the brain does. It is a fun-damental property of any biological living organism. Memories are at the core of being human because they thread and unify our overall sense of being."

This is indeed a much grander description and may feel a little nebu-lous. So let's break memory down into its four processes. Encoding, the first step of memory formation, is all about your senses—the sights, sounds, tastes, smells, and touches that bring together the richness of an ongoing experience into the cohesive whole that becomes a memory. These general sensory categories might feel arbitrary—Can you really separate the *smell* of an onion from its *taste?*—but there is good reason for this "batching" of the Encoding process. It helps scientists find patterns in how the brain *takes in information over time.* The first time I visited my family in El Salvador, when I was about six years old, every moment was entirely new to me, something I had never experienced before. It was like having my senses dialed up to high definition: the warm pebbles tickled my feet when I stepped into the lake near my grandparents' house; the carne asada that my grandma cooked would waft through the rooms of the house, and soon the entire block would smell like a steakhouse; I could see the country's volcanoes from miles away and wonder if they might erupt again, all while hearing roosters crowing and horses galloping by me. My brain was taking in and encoding every bit of sensory detail, putting together my wonderful experience of being in the place where my parents grew up.

The next stage of memory is Storing, a process that begins the moment after an experience is over. In this phase, the brain creates a record of the past by storing information in its own cellular terms. In contrast to Encoding, Storing is the basic way that an experience creates a mappable, observable change in the brain at the cellular level. These changes are stored copies of experiences that allow us to mentally time-travel back to the past.

Let's deconstruct that early memory of El Salvador as an example. The night of my first visit, I went to sleep on my grandma's hammock and was thinking about everything that happened that day. My mind was effortlessly replaying the parts that stood out to me the most—the lake, the pupusas, the mountainous landscape fading into the horizon. Storing a memory doesn't have to be a conscious effort—our brain will do it whether we're aware of the process or not—but it was as if my brain was rehearsing the day's events to help solidify all the details, big and small, into a memory. From the brain's perspective, conscious and subconscious rehearsal makes a memory, and so the brain rehearses its own experience day and night to effectively store it away for later use. And best of all, there's no test at the end of this kind of practice rehearsal—just my brain's ability to store what I considered to be a memorable day.

Once a memory is stored, it is ready to be retrieved when called upon, and the things that memories then do are marvelous. When I retrieve my memory of my first visit to El Salvador, I mentally time-travel back to specific moments and swell with the admiration and curiosity that comes with thinking about my family's deep history. Retrieving a memory means I can "rewind" to a snippet of my past and then zoom in and out to replay the granular details of pebbles or to gaze at a spectacularly star-speckled sky or to relive scenes of breakfast with my family, again and again. Retrieving these kinds of memories is when we begin to find that thing called "fulfillment" and all the moments in which it has been sprinkled throughout our past.

These phases of encoding, storing, and retrieving a memory, however, don't happen in some vacuum. From the brain's perspective, experience never stops, and this means that we're always encoding bits and pieces of this, while storing bits and pieces of that, while retrieving some aspect of our past. If I were to give memory a movie title, it would be *Everything Everywhere All at Once-ish*. And it doesn't end here: the act of recalling a memory leads us to one more magnificent property of memory, and that is its malleability. Memories become updated with new information each time we recall them. This Updating process taps into our "library" of memories and scribbles new information into one of our books. I *know* that there were pupusas on my first trip to El Salvador, but no matter how hard I try to recall what else was on my plate or what clothes I was even wearing at the time, the surrounding details of the meal simply shape-shift with every attempt. Sometimes there's a bit of yellow rice at the edge of the blue ceramic plate, and other times there's a slightly burnt tortilla; sometimes there's a coconut cut open to my right ready for drinking, and other times it's a Cola Champagne. Sometimes I'm wearing a gray Nike T-shirt and other times I'm wearing a purple soccer jersey. All of these details infiltrate my memory and warp it into something new each time—into some version of *my truth* that, like memory, is defined and redefined every time I remember.

Okay. But what does a memory *mean* to you? What does it mean to remember, to forget, to choose what to try to remember with greater clarity, and what to let go of? Do we even have a choice? For example, I'm attempting to remember past events for this book. What challenges does that pose?

No one can claim that they've solved the mysteries of the brain, let alone of memory, but Xu and I saw this as a challenge (even if it took dedicating our lives to getting an answer). We could start with the four main components—Encoding, Storing, Retrieving, Updating—which at least allow us to break this complicated part of cognition down to some of its constituent parts and to tackle each with neuroscience.

———

Six months after Xu and I successfully shot lasers into the mouse's brain and reactivated a fear memory, we submitted our work to the journal *Nature*, and in 2012 our paper was published. We had localized a memory in the brain and had artificially triggered that memory. Our paper used words like *hippocampal dentate gyrus* and *engram* and *Channelrhodopsin-2 (ChR2)*, but the international news coverage referenced movies like *Memento* and *Eternal Sunshine of the Spotless Mind*. Media calls and interviews followed, as people pushed us to contextualize our findings and speculate on the future of memory research.

For once, the international news coverage was apt. Within just three years of our discovery, between 2012 and 2015, the scientific advances in memory control and manipulation in our memory research community were mind-bending: creating false memories, erasing memories, manipulating positive memories in the context of depression, bringing back memories that were once thought to be lost in Alzheimer's, tinkering with memories of social experiences, turning good memories bad and bad memories good.

If all of this sounds too good to be true, that's because it is still wildly incomplete; it is all part of a larger revolution brewing in science to make memory manipulation a commonplace practice in the lab. Even before our paper was published, several labs around the world were working on the same project, in a race to publish first. Egos, funding, awards, and recognition meant navigating a—scientifically speaking—suboptimal future. In other words: it was stressful as shit. A race over bragging rights, and not necessarily the significance of the discoveries or how scientists might combine efforts and work toward a common cause, took over.

You might say that Xu and I were naive. I prefer to say that we were young scientists with far fewer encoded and stored memories of professional slights and nursed reputational injuries to retrieve, and we tried to operate outside of the rat race. For us, turning the process of discovery into a career-long race would kill the part of science that felt most essential and human to us. What we were in practice, however, was frustrated by this competitive aspect of experimental science, especially since we were still learning how to navigate the politics within academia.

Xu gave me some advice, which he would often repeat: When faced with the nastier side of science, don't just put your head down and ignore it. These feelings aren't just something to be brushed off; they're exactly the things we can learn from. They're our mind's way of putting what we value into focus. "Fight fire with water," Xu would say. Fighting fire with fire burns the edifice of academia down; fighting fire with water wins by solving the problem.

Xu's guidance always came in a measured voice, a calm cadence that showed me a diplomatic approach to problem solving, even if the environment bothered him too. He was like that, a scientist firmly grounded by the gravitational pull of the meticulous practice of his craft. I admired his composure. Neither competition nor ego would sway Xu.

Xu and I found ways to manage the competition, but the ad hominem criticisms of our "place" within academia were, for me, difficult to endure—infuriating in fact. Here's a common criticism we heard: "Xu Liu and Steve Ramirez only got the paper accepted in *Nature* because the journal has a minority quota." This kind of statement was nothing new. People spewed similar judgments in my direction all the time: "MIT had a minority quota," "*Forbes* must have needed to include a minority," "TED had a quota," "Harvard *clearly* had to increase the number of Latinos," "NPR had a DEI mandate." I started to wonder if people thought I'd somehow managed to *turn* Latino just to get an edge in academia. And Xu started to wonder if he'd ever be able to talk about his science without someone criticizing *him* as opposed to critiquing his work. He once said to me, "If you think being a minority in science makes it *easier* to get an award or a job, then you don't understand why minorities are, well, not a majority."

In my first few years of graduate school, these kinds of comments made me livid. As did admonishments like, "Don't let them get to you," and "You've got to ignore the haters," and "Just stick to the science." I didn't want to feel like I had to be thick-skinned and bulletproof to survive. Why did I need to dismiss my humanity to do good science? These things *should* hurt, I thought. Desensitizing myself to them made it all too easy to become complacent about the toxicity present in some academic circles. But the relentless nature of the "minority quota" comments had a cumulative effect. Each one did its part in deflating my sense of accomplishment that came with a scientific discovery. My response wasn't to become desensitized but to double down. "I have to discover more" was my ever-present thought, my solution, my mind's way of coping. It's the one thing I felt I could do really well: I knew how to work my ass off.

———

One evening in the not-too-distant past, I was at the top of the Prudential building in Boston listening to a jazz band that played on Wednesdays in an open lounge area surrounded by a floor-to-ceiling panoramic view of the city. It was one of the highest places I could possibly be in Boston, away from the ground-level of reality, and I'd come to this nook of the city alone when I needed time to just *be*. Halfway through the evening, a bouncy bass line emerged from the stage with the kind of weight and rhythm that begins in your belly and radiates outward to tip-tapping feet and tempo-following head nods. The jazz lick was the backbone of Glenn Miller's "In the Mood," a song that gave me so much trouble when trying to learn it on the piano as a kid that I still sometimes hear it in my dreams. I closed my eyes and, after one big sigh, let myself *be*. My left hand began playing the arpeggio on my thigh as each note rose up from my childhood and . . . suddenly . . .

————

Xu and I were at the top of the Prudential and smiling back at Boston's bright lights as they shimmered over the Charles River. We were having a fancy dinner together (all expenses paid for by our boss) to celebrate the publication of our discovery on memory manipulation. I could see my childhood hometown; my alma mater, Boston University; my current apartment by MIT; and Xu's reflection on the window, fifty-two stories in the sky, all in a single field of view. The music filled the lounge with the kind of kinetic cadence that synchronizes everyone in the room.

I've never been so happy and so fully alive.

I knew I'd repeatedly come back to this moment in my mind. The music and Xu's voice were becoming part of my life's soundtrack.

"Are you in the mood for dessert?" I asked, while the jazz band played on.

"I'm a little drunk," he responded, grinning at me. "So . . . yes."

His peppy, vulnerable, and playful confession caught me off guard, but I accepted the moment as one we had earned: a bit of debauchery amid the stress of science.

"Xu, you're not going to believe me, but just ignore the entire menu and trust me here. I bet you my left hippocampus that if we ask the waiter what to order, they'll one thousand percent say the cookies and milk."

"Deal."

When the waiter came over, I asked what their best dessert was, and he simply took our menus, folded them up, and said, "I got you. It'll be about fifteen minutes."

While we were waiting for our dessert, Xu asked, "Does it ever scare you that you can hold your thumb out, close one eye, and see where ninety-nine percent of your life has played out?" He was motioning toward the window.

"I find it weirdly calming," I said.

"It's crazy to think where we'll be in a few years, given how much has happened since we published our paper," Xu said. "It's kind of scary, but I think we'll be okay."

"I think we'll be more than okay. Above average. Let's shoot for above average," I said.

"I like that. Do you want the rest of my above-average drink?" Xu slid his golden ale toward me.

"I got you," I said instinctively, as I traded my empty glass for his full one.

The waiter came over and ceremoniously placed a glossy porcelain dish in the center of our table. A dozen freshly baked cookies were stacked on top of each other—some oozing chocolate chips, others smelling of nothing but pure butter and sugar, some with crumbly oatmeal and raisins, and others infused with a velvety peanut butter. Berries lined the dish, powdered sugar dusted the top of the stack, and a side of French vanilla Chantilly cream tied it all together. This was served with a cup of cold milk because what else does one wash down the Platonic ideal of cookies with anyway?

"Enjoy," the waiter said, while a handful of people in the lounge looked over to see what on earth smelled like heaven.

"I didn't want your left hippocampus anyway," Xu remarked.

"After you!" I declared, anticipating Xu's reaction to his first bite. He picked up a chocolate chip cookie that began folding in on itself, dipped it quickly into the cream, and crunched off half of it in one go. He nodded approvingly . . . and fixed his glasses.

It was a childlike moment, paired with our adult scientific achievements, one that I could store in my memory bank in the "Feel Good" folder.

Xu started to laugh, perhaps realizing that the cookies and milk were as damn good as I'd claimed they would be. The music came into the foreground with a lone bass line . . . which brought the room back to baseline.

"Last call!"

———

My left hand finished playing the arpeggio on my thigh, as each note brought me back to where and when I was. I was alone. Boston glimmered in the background.

I realized in that moment, wishing Xu were sitting across from me again, that the power of encoding, storing, retrieving, and even updating an ordinary memory was changing me as a person. I was at once becoming more aware, frightened, and reverent of what memory was doing *to me* because the force of memory lasts well beyond its wake. Sometimes I'd recall this memory with Xu and feel empowered to continue doing science. I'd emerge from the memory having grown a bit in my confidence to navigate science and the world without my friend. There are always more cookies and milk to be shared in life.

Other times, this same memory led me to imagine all that could never be with Xu. I'd feel overwhelming sadness because it was a moment I could never recreate with him. Instead of trading my empty glass with Xu's full one, I'd trade my empty glass with the bartender.

And sometimes I'd remember for the sake of learning something new. I'd go to the top of the Prudential to get lost in the music and memories until they all brought Xu back. And when they did, I would say the things I wish I'd said to him years earlier. I would hug him for longer than I ever had before. And I would order dessert again to see him laugh out of pure joy. Retrieving this memory was a miniature miracle in its ability to bring the past back to life, so I'd go to the lounge to relive this moment with my memory of him again and again. Each time felt full of purpose—I was learning to rewrite what my past meant to me.

Just what *exactly* that purpose is continues to change and will evolve throughout the narrative in this book. All I know is that learning about memory brought me to Xu, one of the closest friends I'd make in my life. Memories now bring me back to him. In a way that I find biologically meaningful, Xu and I are still connected. I sometimes even dream about our time at the Prudential together and relive it with fantastical details. The music plays the same, but the skyscape behind the lounge shape-shifts into the buildings on MIT's campus or some of the playgrounds from my childhood. Retrieving a memory offers a chance to change what it means to us, to update it, regardless of whether we're even awake. The meaning of our memory together at the Prudential is mercurial, but it's where the *should haves* of my life with Xu find resolution. This is comforting because a memory may transform me entirely, but I have the power to transform it as well—both with my mind and with my science.

At its heart, *How to Change a Memory* is about these transformations: your brain transforms moments into engrams, the theorized physical units of memory imprinted in your brain, and these engrams are controllable—they can be transformed and in turn can transform us as human beings. That memory has so much transformational power is the most astonishing thing that I know, and I have dedicated my career to understanding just how far we can push this power in the lab. In this book, I will take you on the journey that I myself traveled, both professionally and personally, to understand the new science of memory manipulation. I've started the story of that journey in this chapter, with my and Xu's breakthrough in artificially activating a memory in a rodent brain. In many ways this experiment—and our friendship—was indeed the breakthrough that jump-started my career and propelled me forward into a new chapter in my life. But like all stories, this story has deep roots, a backstory that did not start with me, or with Xu; our research in memory manipulation rests on a foundation of contemporary triumphs in neuroscience, which we explore in part I.

Twenty-first-century neuroscience has successfully tracked where memories reside (chapter 1), and it has measured how they change over time (chapter 2). This work has permitted neuroscientists to erase and activate memories of all kinds (chapter 3), as well as to create and implant false memories (chapter 4), all with extraordinary clinical applications (chapter 5). We'll follow the path of some incredible scientists whose work has brought us to the very cutting edge of memory research. We'll unpack their breakthroughs and build a new understanding of how memory works, as well as how we can begin to artificially change memories to enable biological well-being. The goal of part I, therefore, is to provide a framework for the neuroscientific basis of memory and its manipulation.

In part II, we look at how artificially controlling memory changes our very understanding of the nature of what memory is for and how it intersects with our lives. Memories build our futures through dreams and imagination (chapter 6), and they sculpt our overall sense of being to endow us with an identity (chapter 7). Manipulating memory has the power to change life as we know it: our quest concludes in chapter 8 with a forward-thinking account of the promises and perils of memory manipulation. The goal of part II, therefore, is to illuminate just how intertwined the human condition is with memory so that we emerge with a profound appreciation for the very thing that enables our livelihood and sense of self.

My quest to manipulate a memory is as scientific as it is personal. In each chapter I fold in some of the most valuable memories I've made in my life, both in and out of the lab. These personal stories are meant to showcase how the scientific process works from the perspective of the people who did the discovering. There's a very human element at play in every discovery, which is sometimes overlooked or not told at all. Just as the work that Xu and I have done to manipulate memory builds on the foundational neuroscience that came before us, my own personal journey as a scientist is the product of my past, of my own memories, and of memories that predate me. My parents and siblings came to the United States from El Salvador, fleeing civil war, many years before I was born; their memories and their love for me form the foundation of who I am as a scientist.

I've struggled along the way, as we all do, but I've also experienced the joy of true friendship and the fulfillment that comes from making my parents proud, both of which have imbued my work with meaning. I've learned so much about how memories can outlive us and the heartbreak that occurs when the things we love become memories themselves. There is a pulse to memory that beats on, far beyond the biological lifespan of the brain which housed it first.

My quest to manipulate a memory began with a friendship. Xu and I belonged to each other.[1] The truth is that we all contain engrams of each other—engrams of those who are living and engrams that outlive those who are gone. This book is my attempt to make sense of the enigma of memory—the snippets of remembrances, the brief moments in time, the decisions we make, the blackouts, the imagined, and the dreamt of—all the things the brain does to breathe life into the past so that we can heal and become whole again. This book is my engram of Xu, and how my engram managed to outlive him, too.

Part I

1

The Hundred-Year Quest
for the Engram

MY QUEST to manipulate a memory brought me to Xu. The first time I met him, he was hunched over his lab bench with a red pen in hand, meticulously writing labels on opaque test tubes the size of Tylenol pills. *ChR2-GFP* on one, *Rabbit anti-c-Fos* on another, *488 anti-Rabbit* on a third. He had clearly done this a thousand times and looked more like a machine automatically applying labels than a human focusing on the importance of the chemical at hand.

I was only a few months into my first year of MIT's Brain and Cognitive Sciences graduate program and didn't quite know what to make of the scene in front of me, so I decided to simply take it all in. Off the long hallway of the lab, at right angles to it, were a dozen long lab workbenches, each with enough space for two researchers to sit at while the rest of the lab worked in adjacent rooms. Coffee-stained papers were scattered across the desks situated at the end of some of the benches; extra-large binders, overstuffed with notes spanning years of experiments, were precariously stacked on one desk nearby, a heap of workout clothes piled underneath. Partially filled test tubes were lined up on some benches, to be used for an ongoing project I assumed, and bits and pieces of half-put-together electronics were strewn about on yet more benches, perhaps to build new computers. Sterilizers and mini-centrifuges were humming and buzzing in the background, set up to get the day's tools and chemical reagents ready to go. Organized chaos, I thought, minus the organized part.

Except Xu's area was an oasis that hinted at what "organized" might mean.

Xu's lab bench was pristine. There was a shelf above his computer, which held about a dozen black binders, each labeled with the year and the experimental design they contained. Post-it notes were affixed to the shelf, a careful line of overlapping To-Do lists with everything checked off. His test tubes, pipettes, surgical equipment, and chemicals all had a dedicated spot on the bench, and he had labeled each item with color-coded lab tape. It was like a picture-perfect supermarket for science: aisle Xu was stocked and kept tidy, while aisle rest-of-the-lab looked like the employees were on strike.

Xu's organization was intimidating, but something about how he approached the day-to-day grind in science felt welcoming, which was important to me because I was still feeling out the people and the projects in the lab.

During the first year of graduate school, most students go through "rotations," spending a couple of months in each of several labs to find a suitable mentor, learning new strategies for doing science, and getting to know the lab members and the culture. It's a time to see which lab is the best fit for the subsequent eternity that is also known as "getting a PhD."

My undergraduate research advisor had recommended that I rotate in the lab of the Nobel Prize–winner Susumu Tonegawa, whose work focused on studying the biological basis of memory. My advisor had devoured work from the Tonegawa Lab for decades and gave me an overview of how the lab was connecting everything from molecules and cells to cognition and behavior. It was a lab in which attempting any kind of project in neuroscience seemed to be fair game. I thought of it as an open-ended, choose-your-own-adventure science story, with a little bit of "my career is on the line" added for good measure.

Early on in my time in his lab, Susumu encouraged me to chat freely with everyone there. He told me to see if there was any project that stood out as something I could build on for my PhD. At the time, there were almost sixty people in the lab—most labs contain less than ten researchers. I was too intimidated to reach out to anybody, let alone Xu. Instead, I spent the first three months sitting at a desk thinking, *Well, shit, now what?*

Eventually a few of the graduate students in the lab started taking me out for coffee. They were trying to be friendly, but they also wanted to give me a warning: people tended to stay in the Tonegawa Lab for a *very* long time before graduating, they told me, often because the projects are on such a large scale and with such new technology that troubleshooting each step took years. At the time, though, I didn't care how long it would take me to graduate; I just wanted to be trained as a competent neuroscientist.

Feeling emboldened by the support of the other students, I went back to Susumu and asked for his help finding a project. He looked at me with impish eyes, and told me to stop by his office at the end of the day.

A few hours later, I made my way down the corridor, the lab's publications in paper form lining the walls, and arrived at Susumu's office. It was a long, narrow space, packed with books and journals on shelves that went up to the ceiling on one wall, science papers that were neatly spread out all over his desk at the end of the room, and about a half-century's worth of awards sitting atop any available desk space. I could barely see his face behind the computer monitor, which he was staring at with an unflinching intensity.

He invited me in with a wave, motioned for me to sit down, and got down to business. Susumu described the dizzying number of projects going on in his lab, over twenty at the time, and seemed particularly animated about one that involved combining the lab's newest technology to study how memory works. This project sounded like the perfect blend for me of science fiction and "this will never work." Sensing our mutual excitement, I asked Susumu if I could learn everything about the project. I like to think that this single question sparked a connection in Susumu brain's that would soon bring Xu and me together.

I noticed that Susumu kept popping back behind the oversized computer screen to steal a glimpse at something. I assumed it was data from the recent experiment on the hippocampus and memory. I glanced over at a picture of Susumu celebrating the opening of our neuroscience building, a decade prior, and then to a row of windows on the opposite end of his office where I could see the dizzying architecture of some of MIT's buildings. I was nervous, trying not to fidget too much, and sank myself into his ridiculously comfortable leather couch.

He looked over at his monitor again, and his scrunched-up face revealed it all to me: as a fellow sports fanatic, I realized he *had* to be checking the score of the Red Sox game.

I jumped at the chance to connect with him. "I used to go to games a lot with my dad," I blurted out, "and since I was born in Boston, I think I'm contractually obligated to be a Red Sox fan."

Susumu chuckled. "Oh! Well, a guy I know gives me tickets to some great seats every season, so let's go to an upcoming game with a few other people in the lab," he said in his matter-of-fact style.

He promptly rose from his seat and nodded toward the door—my cue that we'd continue our more academic talk on our way back to the lab to find some fellow Sox fans.

As we walked to the lab, we talked more about the ongoing projects.

"It's amazing how far technology has advanced in neuroscience," he said, "and yet we still don't know how something we use every single day works." He was referring to memory. "For two decades the lab has been able to delete genes to see how their absence affects the activity of brain cells and how the now defective brain cells impair memory. We've also been able to inactivate the communication between very specific brain areas to break down the processes required for memory. But memory still feels as mysterious as ever. I remember throwing out the first pitch at a Red Sox game a few years ago, and I can describe the entire thing to you—but I wouldn't be able to tell you how my brain is *really* doing this."

Most of our conversations, it soon occurred to me, would focus on science or Boston sports—quite literally the only two topics we ever talked about at length. Luckily for me, I happened to love both.

Susumu brought me first to Xu's desk.

Susumu tapped Xu on the shoulder, breaking his hyperfocus, and asked if he'd like to see a game sometime soon. Xu happily took Susumu up on his offer, excited that the tickets for the game included fancy private seats with unlimited food and drinks. Then Susumu formally introduced Xu as "a postdoc doing some very exciting science."

Judging by Xu's intense level of engagement with his science, not to mention his immaculate lab bench, I thought, *That's probably the understatement of the year.*

"My research is *something* anyway," Xu responded. His response was noncommittal on the surface, but I could feel the optimism beneath it all.

I have since blamed Xu's nonreply reply for my faux pas—I said, "Hey, ready for some baseball!?" and inwardly face-palmed myself, realizing that my informal and overly enthusiastic greeting might be off-putting to someone in the middle of trying to make, oh, I don't know, *memory manipulation* a freaking thing.

I like to think Susumu face-palmed himself, too, as he immediately interjected and asked Xu to explain what he was working on—and then wished us good luck and left us to our own devices.

Xu immediately launched into teaching me how to mix a set of chemicals necessary to stain brain tissue so that certain brain cells would be tricked into glowing a vibrant green under a microscope. He spoke slowly, with purpose, as he described terms that meant nothing to me at the time. *Doxycycline. Tetracycline transactivator. Activity-dependent c-Fos upregulation. Adeno-associated viruses.* I tried my best not to give off that deer-in-the-headlights look. Xu's patient lecture continued. He took a small slice of a lab mouse brain and transferred it to the middle of a thin, circular dish filled with saline for me to view.

"If you look closely at the two shapes that kind of look like the infinity symbol or a thin pair of glasses, that's the hippocampus. We've been trying to make a genetic tool that lets us see which brain cells hold onto a memory here. If it works, we'll be able to see those brain cells under a microscope because the tool forces only cells that were involved in making a memory to glow green. So we should be able to take this brain slice through a staining protocol that will enhance the green color in a way that makes it easily detectable."

I returned the dish, and Xu paused to place it carefully on his lab desk.

"I feel good about this one," he said, "but I've been saying that for a few years straight now."

"What does c-Fos do?" I asked. "Why do we need doxycycline? Are the viruses safe to use? How long does this all take?"

Xu paused before responding to each of my many questions, seemingly thinking through his answers word for word. The small, inaudible movements of his lips indicated his stream of consciousness becoming organized thoughts. When he was ready, each of his measured responses began with reassurance: "That's a really good question." His encouraging manner helped bring me into his world of c-Fos and viruses, as he adjusted his clear-rimmed glasses and sat upright, his black Chuck Taylors tapping on one of the chair's legs underneath him every time he got excited to dive into more and more detail with me. It took only this first interaction with him for the entire experience to be encoded in my memory as something tremendously meaningful to me: he made me feel *valued*.

"If the cells glow green then we're in business. The first big goal here is to find out which brain cells hold onto a memory," Xu explained. "To do this, we have to be able to see which ones are 'on' as a memory is being formed. If we can do this, then we're on to the second big goal, which is to activate them. If we see green, then we'll know which cells to target if we want to artificially turn them back on and test if this makes the memory come back. That's where our genetic tool comes in because it can help us identify these cells first so that we can then trick them back on."

Before I got too comfortable, he added, "Here is a list of papers that should get you up to speed. You have to devour the literature so that you can come to know what you don't know."

Without even a pause, Xu told me not to be nervous and to slap on some gloves and dive right in.

I watched as Xu used a thin paintbrush to carefully transfer mouse brain slices, each the size of a penny, from one circular dish filled with saline to another. These brain slices contained key information on whether the genetic strategy that Xu was using was sensitive enough to detect which brain cells held onto a memory. At the time, though, all I saw were tiny petals of an opaque white brain that Xu considered to be important.

Xu was meticulous. He was quiet. And he didn't want to be interrupted. I sensed that he had a blueprint in his mind of all the motions he had to perform to make sure an experiment went exactly as planned. I hurried to put on a pair of lab gloves and, in my rush, my hand ripped right through the first one. My scientific career at MIT began with strips of latex dangling from my hand.

After a successful second attempt at being a non-embarrassment of a human being, I grabbed a paintbrush to begin learning how to delicately handle these 50-micron-thick slices. Jostling them with the brush felt like using a fork to pick up a single soggy piece of corn flake, so Xu gave me some extra brain slices to move around just to get a feel for them. Gliding the brain through the saline too quickly would shred it; if I was too slow, I'd also get nowhere. Every few minutes Xu would stop to teach me how to move my wrists to roll a brain slice onto the paintbrush, and then unroll it onto a microscope slide without causing any damage. It wasn't exactly the Sistine Chapel, but I did feel like I had found myself in the middle of some grand canvas of memory research.

Please, Steve, for the love of God, don't fuck this up, I kept thinking.

I'd been having this same thought repeatedly since I arrived at MIT. I had only the vaguest idea of what I was doing when I applied to graduate school. It took me most of my time in college to figure out that I wanted to pursue a PhD in science. I loved reading and rereading Shakespeare and thought maybe I could become a scholar of his works. I had been playing the piano since I was six years old and could see myself playing music forever, but I turned down a music scholarship to attend college without a major because it meant I got to live with my two childhood best friends. I watched *House* weekly and romanticized the possibility of practicing medicine. I wondered what it would be like to train as an astronaut and float in the International Space Station. Or maybe I could win a competition on the Food Network and open up a bar called La Sagrada Sangria. Or maybe I could work in a lab and make discoveries about our inner biology and share those discoveries with the world.

I reasoned the only way I could begin ruling out possible career options was by getting firsthand experience. Science seemed the best place to start, given that all my interests involved *creating* something new. So in my sophomore year at Boston University, while working at a local Walgreens, I started cold-emailing biology professors during my fifteen-minute breaks, offering to volunteer in their labs. Only one professor responded, and I ended up working in his lab for a summer, mindlessly mixing the same chemicals all day, with no real sense of why I was doing so. But I did know one thing: I was sure as hell not creating anything new. Quite the opposite. I was literally following the same recipe to make the same liquid to store biological tissue in the same way every single day. "Fulfilling" this was not. This couldn't possibly be all there was to science, right?

A friend pointed me in the direction of a professor who she thought might help. Over the course of several conversations, Paul Lipton (who would soon become the director of BU's undergraduate neuroscience program) planted a powerful idea in my mind: Why not study the organ that's created everything that's ever been produced? *Hamlet*, the Fifth Symphony, vaccines, rocket ships, sangria—all these achievements were the product of the most multidisciplinary organ known to humankind: the brain.

Halfway through my time in college—still a couple of years before I set foot on MIT's campus—Paul introduced me to my eventual research advisor, Howard Eichenbaum, a towering force in the neuroscience community and, like Susumu, an avid Red Sox fan. Howard invited me into his lab, where everyone had an unmistakable passion for studying memory. Each lab member would sit down with me to explain their projects in detail, before inviting me out to play pool or darts in the evenings. They made me feel comfortable working by their sides, despite me being an amateur neuroscientist.

I worked in the Eichenbaum lab until the end of my senior year. Even though I didn't know for sure that I wanted to do science forever and become a professor, the brain was still a mystery to me, and studying it with a team of researchers felt like the kind of journey that I didn't want to end anytime soon. The togetherness of doing science had me hooked.

The year 2010 found me at the end of my senior year of college, eagerly waiting to hear back from my applications to various neuroscience PhD programs. My seventeenth-floor dorm room at BU overlooked the famous Great Dome at MIT. I'd go to bed gazing at it—sometimes with hope, often with envy. It was my Gatsbian green light: a dream I could stretch my arms out to but never quite touch.

Howard had convinced me that I would be bananas not to apply to MIT. He knew that I busted my ass off in lab daily and that I came in nearly every free hour I had to get the research experience I needed to feel ready for graduate school. Admittedly, it was daunting to learn that weekends and holidays sometimes don't exist for researchers—mice and rats don't celebrate Christmas or Cinco de Mayo, and experimental schedules can last for months without pause. Still, I was determined to show up, do my job, and give science a shot.

Going to graduate school, he told me, was all about reaching the edge of what a field knows—and then advancing the field through the arduous process of discovery. It would be a time for me to learn how to test hypotheses about how the world works and to form a cohesive narrative around what I observed. It was also a necessary step on the path toward a myriad of job opportunities, including a career in academia like Howard's. Whatever was to become of my career, I knew that I wanted intellectual freedom and to be my own boss.

However, knowing what you want and imagining it possible, even plausible, are two very different things. The more Howard talked of MIT, the more I thought *Howard* was the bananas one for pushing me to apply. But Howard and his team won out in the end.

———

Then came March 15, 2010. I began the morning curled up next to a toilet, suffering from a colossal bout of food poisoning. I couldn't hold anything down, let alone listen to a lecture on the molecular biology of the cell, so I skipped my classes that day and played *Guitar Hero* when I wasn't napping or dry-heaving my life away. That evening, hollowed out but finally steady on my feet, I went to the dining hall alone and filled my bowl with milk and a mountain of Lucky Charms and Cap'n Crunch—*so cool*—while my roommates studied for finals. I checked my phone.

> Dear Mr. Ramirez,
>
> On behalf of the Admissions Committee, I am pleased to offer you admission as an entering graduate student in the Department of Brain and Cognitive Sciences at the Massachusetts Institute of Technology, effective with the beginning of the fall term in September 2010.

LET'S GO! I called my parents to share the news.

"Dad!" I screamed—right before my phone fell out of my hands and splashed into my milk and cereal. The latter was a splashed mess; the former was dead because phones weren't yet milk-proof. I ran back to my dorm to borrow my roommate's phone and, after reassuring both of my parents that I was still alive, I told them the news.

"*Estamos muy, muy orgullosos de ti y de todo lo que has hecho,*" my parents said over and over. *We are so, so proud of you and everything you've done.* I could feel a sense of fulfilled dreams swelling in their voices.

My parents never finished high school.

My dad snuck into the United States—"No human being is illegal," my parents would say—in the early 1980s to escape from the civil war in El Salvador. Then he had to earn enough money to bring my brother, sister, and mom to the States. This often meant doing the jobs that no one else wanted to do: his first job was going door-to-door to convince the living to buy tombstones for when they're . . . not living. His second job was working the night shift as a janitor at a local brunch spot, cleaning up the scrambled eggs and beer that would line the floors every evening. And his third job was working the day shift as an animal technician at an animal locomotion lab at Harvard.

To highlight the grit my dad had when it came to getting shit done: within a few years of selling tombstones, he was a manager; at the end of his first year working at the brunch spot, he became the head chef;

and after a few years of working in the lab, he became the lab manager, which was a title he sustained for over thirty years.

I mean it when I say I *never* heard my dad complain, despite balancing three jobs and hundred-hour shifts every week for years. To the contrary, on his one day off, we'd head to the theater and see how many movies we could sneak into together. When he had to work back-to-back shifts, he'd take me with him to the brunch spot to get an ice cream sundae, and one of his close friends would drive me back home. When he'd take me to lab, we'd turn janitorial work into a game: if I could mop all the long hallways without missing a single spot, we'd get Burger King on the way home. In retrospect, perhaps Burger King was always part of the plan.

Despite his schedule, every Monday I'd go with my dad to a sports complex north of Boston, where my cousin and I would play in an adjacent arcade while my dad captained his soccer squad, often playing in games that went past midnight. During school holidays, he'd take me to work with him, and then we'd go fishing and get milkshakes together. On top of it all, every night for as long as I can remember, my dad and I would wind down together with the same routine: he would pray for a few minutes while I closed my eyes next to him and rehearsed a prayer or two, followed by mentally revisiting all the moments of the day for which I was grateful. His indefatigable optimism made me feel like all the world's problems were solvable.

The only thing my parents asked of me was: get an education and help others when you can. A common saying in my family was, "It's not a matter of *if* I can do it, it's a matter of *when and how* I will do it." For me, MIT was the *if* that I had just turned into a certainty, and it was now my job to use it for some kind of good.

So like my siblings ahead of me, I was now going to get an advanced academic degree. On top of it all, I didn't have to worry about financial aid and taking out more loans because, this time, my dream school was going to pay *me* to get my PhD. *That's* what's up.

Deep inside, I damn well knew I'd earned it. I didn't just accidently sneeze and wake up with an acceptance letter. I'd put in hours and hours in the lab. Still, my doubts would surface routinely. For starters, I was in the privileged position of having access to a kind of education my parents never had, and it was hard not to feel like I got stupidly lucky, like I didn't really deserve to be at MIT. And there was something else: Was it really *me* that got into MIT or did the admissions committee see me as a box that could be checked: Latino + coming from the Eichenbaum Lab = good for their numbers. Or maybe someone else that they admitted first ended up going to a different school and I was the second choice?

It may take a village to raise a child, but I believe it takes the love of a parent to empower one. My GPA was good, but it was what we in the business call "not a 4.0." And, according to my standardized test scores, either I was lousy at standardized tests or I could barely string together a subject and a verb or do long division and carry the one. But I knew I wanted to study the neuroscience of memory, which required me to ask questions and do experiments in the lab. My parents reminded me that this was what I was all about. "Remember what your piano teacher told you when you were little, when you asked him what was the secret to playing 'Für Elise' because you kept struggling with it so much? The secret was to learn it *until your fingers remember*. Let Beethoven become a part of you." Maybe science is the same. Ask questions, do experiments, mess up, and keep on messing up until one day it'll click, and all that trying turns into knowing how to play. Let science become a part of you.

Their perspective certainly helped ease my nerves, but that feeling that you're not good enough is *really* difficult to escape, admittedly, and it sometimes slaps you in the face when you least expect it.

The first time I met the other graduate students in my class, we did one of those awkward ice breakers where you sit in a circle and share a fun fact about yourself. Most of the people in my group talked about how they had already published their scientific discoveries in prestigious journals as undergraduates. Meanwhile, my fun fact was about how, in high school, I had once eaten a two-pound steak called the "Bad Larry" that the Texas Roadhouse had on their menu as a challenge to anyone foolish enough to order it. The affable graduate student next to me shared that he was a cognitive neuroscientist. And was still a teenager.

Are you serious? I thought in disbelief. As a teenager I was trying to figure out how much steak I could force myself to eat without feeling like a beached whale, and this prodigy was already deep into high-level computational neuroscience on how the brain makes sense of the visual world around us.

Comparison, I would learn, is indeed the thief of joy.

Before meeting Xu, my family's shared memories were my main source of defense against feeling like an outsider looking into a community I so desperately wanted to join.

Whenever I felt lost at MIT, I remembered my parents working hundred-hour weeks to support my siblings and me. I remembered my brother and sister learning English and studying throughout college and law school to become lawyers. I thought of what life was like for my family in El Salvador, where my dad was once kidnapped at gunpoint by a group of army members because he had a beard, which was interpreted to mean that he was likely a guerrilla. The only reason he made it out alive was because he had helped one of the military personnel by giving him food when they were kids in school together, and the guy *happened to recognize him* before someone pulled the trigger. I thought of what it was like for my dad to hug my mom and my siblings goodbye on the day he left El Salvador, knowing that he would not see them again for years—if ever. Adversity is a fact of life, and we all endure our own version: a bad test grade, a rejection letter, a breakup, a career-changing decision, a family loss. All of these will hurt. My parents were my source of composure. So when I felt alone, I could always call my parents; and they answered every single time. Their memories gave me all I needed to feel, at the very least, like I could endure.

Xu's support would end up being just as unconditional and essential.

––––––

"How will we know if the experiment worked?" I asked Xu after transferring over the last brain slices.

He responded: "Tomorrow when we go into the microscope room, we'll look closely at each slice and either see a whole lot of nothing, meaning we have a lot of troubleshooting to do, or something in the brain will be glowing, in which case . . . things get interesting."

After working together that first afternoon for a few hours, we took a break to go to our departmental social, which was every Thursday in our building's common atrium. It is well-known in academia that these events mean one important thing: free food. Each Thursday, we would all hover around the tables, demolishing crackers and cheese or, if the universe smiled on us that day, pizza.

As we stuffed ourselves, Xu and I got to know each other. On the surface, we were different in many ways. Xu was from China, and he loved hiking, seafood, and molecular biology; I was a Salvadoran American who loved sports, steak frites, and playing the piano. He preferred camping over glamping; I preferred to be comfortably sheltered on planet Earth. He would take his time with a beer; I would drink mine while thinking of which kind to order next. He was laser-focused and only ever had a single computer tab open, usually linked to an experiment he was conducting; I was the kind of person who had twenty tabs open during any given experiment and played with my phone while waiting for lab results.

But it was in the seemingly random moments that were more about our past and less about science that we found common ground. As teenagers, we both loved video games. On the weekends when he was in high school back in Shanghai, Xu would sit with his sister and teach her how to play the role-playing game *Diablo*. I would call my cousin to come over and play the newest version of *The Legend of Zelda*. Playing video games was a way for both of us to bond with family while journeying into a fantasy world.

I learned a lot about Xu that day, but I also got the sense that he guarded his privacy closely; the memories he shared with me were *carefully* shared—they felt more like movie trailers—and the glimpses into his past kept me wanting more.

It's all good though, I thought. *This is our day one together, and I'm definitely annoying him by trying to go from zero to best friends in a single conversation.*

Right after the social ended, Xu jumped back into telling me about the goal of artificially reactivating a memory. He did so with such ease and fluidity, compared to just moments before when we'd chatted with other members in the department about all the bewildering science happening around us. I wasn't in any rush to go home and had asked Xu if he could walk me through Project X. The reality was that I noticed how much more conversational Xu was when talking about science, and I didn't want to let an opportunity pass by for us to chat more. We sat down in a sunlit corner of the atrium, overlooking railroad tracks, and every hour when the commuter rail rumbled by, the building rattled with a barely audible hum.

Xu had his laptop with him, of course, and started explaining why it made sense to try to reactivate a memory by stimulating cells in the hippocampus, the area of the brain that plays a fundamental role in producing the memories we personally form and remember. All mammals have a hippocampus, and scientists believe that it's one of the most important areas of the brain involved in turning short-term memories into longer, even permanent, memories. All humans have two hippocampi (don't forget to stick your pinky out when saying this, and while you're at it, remind people that the plural of *thesaurus* is *thesauri*), each tucked inwardly just behind the ears. Brain-imaging studies show that the hippocampus lights up when humans are forming and recalling memories.

Decades of research have established that the hippocampus is important for recalling memories of personally experienced events, such as visiting your childhood home. But memory is also more than just an active hippocampus: for instance, many lines of research have shown that motor areas in the brain are activated when you think about all the motions necessary to walk through your home. The hippocampus lights up with activity when a person imagines navigating the home where they grew up—entering the living room, turning in one direction to find the bathroom, or going in another direction to find the bedroom. Motor areas, on the other hand, light up with activity when we imagine performing physical tasks (which are types of memories more commonly referred to as motor or muscle memory), such as pouring milk into the cereal bowl for breakfast or climbing over a fence during a game of tag.

To show just how powerful eavesdropping on the brain can be with this basic science in hand, neuroscientist Adrian Owen designed a remarkable set of tests to measure whether or not patients in a vegetative state were aware of the world around them. While the patients were in an fMRI (functional magnetic resonance imaging) machine that recorded their brain activity, Owen asked them a series of questions that were supposed to tap into either hippocampal machinery or motor machinery: "Imagine walking through your home"; "Imagine swinging a tennis racket." Sure enough, the hippocampus lit up when a patient was asked to imagine themselves navigating an environment, and the motor cortex lit up when a patient was asked to mentally perform a skilled movement.

The brains of Owen's patients were *responsive*. Something was getting through. He and his group were detecting signs of communication with patients who were, up until that point, thought to be fully unaware of their surroundings.

If patients could use their neural activity to respond to two scenarios despite not displaying any obvious behavioral outputs, then scientists could use this as a kind of language to communicate. Owen and his group asked patients yes-or-no questions; if the answer was yes, he asked them to *imagine* walking through their childhood home, and if the answer was no, he had them *imagine* swinging a tennis racket. Amazingly, some patients were able to answer these questions with their brain activity alone! That's all in the human brain, and it's exactly this kind of trailblazing research that Xu and I devoured in our quest to understand and manipulate memory.

———

Xu's casual generosity on the day we met was very much in keeping with his character. He was the rare scientist who was confident enough in his own scientific abilities to not feel competitive with others in or out of the lab. His unusually open approach to collaborating and sharing credit would become our team's guiding principle. Collaboration was the true heart of our friendship.

That said: the next day, I woke up with an email from Xu that had exactly sixty PDF attachments in it. Some were publications that had come out just months before, on the genetics of memory, and some were documents that were originally written on papyrus. Xu put them in chronological order and told me that the best way to start my education on engrams—on the biochemical changes in the brain that make a memory possible—was to begin with what many consider to be science's original form: philosophy. In the *Theaetetus*, Plato put forth the idea that memory had bodily underpinnings that were analogous to a wax tablet. Using the character of Socrates as his mouthpiece, Plato wrote:

SOCRATES: Please assume, then, for the sake of argument, that there is in our souls a block of wax, in one case larger, in another smaller, in one case the wax is purer, in another more impure and harder, in some cases softer.

THEAETETUS: I assume all that.

SOCRATES: Let us, then, say that this is the gift of Memory, the mother of the Muses, and that whenever we wish to remember anything we see or hear or think of in our own minds, we hold this wax under the perceptions and thoughts and imprint them upon it, just as we make impressions from seal rings; and whatever is imprinted we remember and know as long as its image lasts, but whatever is rubbed out or cannot be imprinted we forget and do not know.

In other words, Plato explained that experience leaves behind a physical imprint in the mind, and this imprint reappears during the process of recollection. However, this is the twenty-first century, folks, so let's update Plato and use a more modern metaphor:

SOCRASTEVE: Please assume, then, for the sake of science, that there is in our brains a pattern of neural activity, in one case specific to learning, in another specific to recollection, in one case the pattern changes a neuron's physical structure, in another it changes a neuron's function, in most cases both.

THEAESTEVE: As a neuroscientist, I assume none of that, but go on.

SOCRASTEVE: Let us, then, say that this is the gift of Memory, the mother of our identity, and that whenever we wish to remember anything we see or hear or think of in our own minds, we experience this neural activity as perceptions and thoughts, which are imprinted upon the brain, just as we make impressions by stitching together pictures, posting them online, and filtering them repeatedly; and whatever is imprinted we remember and know as long as its neural correlates last, but whatever is deleted or does not leave an imprint, we forget and do not know.

In other words, memories are physical, touchable, viewable changes in the brain.

In the last two thousand years, our understanding of the physical basis of memory has gone from metaphors about imprints on wax tablets to microscopic images of what a memory looks like in the brain. The transmutation of thoughts from wax to neurons began with one of the most prominent rivalries of all time. Personal rivalries aren't new: Biggie or Tupac? Taylor Swift or all the Taylor Swift haters? Drake or Kendrick? Messi or Ronaldo? Neuroscience in the late 1800s and early 1900s had a rivalry of its own. The Spanish neuroanatomist Santiago Ramón y Cajal and the Italian physician Camillo Golgi found themselves pitted against each other in a quest to outline the physical landscape of the brain. Their competitive spirits were so fierce that, upon receiving and sharing the Nobel Prize for their work on the structure of the nervous system, Golgi, in his acceptance speech, wasted little time in taking jabs at Cajal and his work: "While I admire the brilliancy of the doctrine which is a worthy product of the high intellect of my illustrious Spanish colleague, I cannot agree with him on some points of an anatomical nature which are, for the theory, of fundamental importance."

Boom, roasted. And I thought competition in *modern* neuroscience was bad.

Golgi believed that the brain was a continuous web of cells, all physically interlinked and relaying information from brain cell to brain cell through direct connections. The idea that brain cells formed a continuous network—or a *reticulum*—was known as the reticular theory. Cajal, however, believed that there was space between neurons and that information was relayed through messengers (chemicals) in that gap.

Was the brain more like a continuous strand of spaghetti intertwined on itself, or like a bowl of separate pieces of rigatoni? To answer this question, Golgi developed a technique of staining brain cells, called the Golgi stain, which made transparent, pink-ish brain cells appear a deep black under a microscope with connections more clearly visible. The process made visualizing a brain cell's complex structure easier. When Golgi used his stain to look at neurons under a microscope, what he saw was a random and magnificent mesh of interconnected neurons. The brain appeared to be a physically connected network of spaghetti.

After learning about the Golgi stain, Cajal went on to improve the method but saw something dramatically different: while the brain indeed appeared to be a mesh of neurons, there were hints of spaces between each brain cell. Cajal painstakingly drew what he observed and produced some of the most stunningly intricate images that exist in neuroscience.

Like most rivalries, theirs was a symbiotic one: Cajal needed Golgi's technique to advance his theories, and Golgi needed Cajal to iterate on the Golgi stain in ways that made the nervous system easier to visualize. But they certainly weren't friends. To them, neuroscience was a contact sport.

It would take decades before more advanced microscopes could take pictures of the nervous system with higher resolution, but an answer eventually emerged: the brain consists of individual processing units that are not continuous with one another but independent functional cells that communicate through chemical messengers that cross small gaps known as synapses. This is known as the neuron doctrine, and it's the basis of every thought and feeling and sensation you've experienced. Every time you stub your toe, watch your favorite Netflix show, or hum a catchy song, individual brain cells work together to process these experiences and bring them to life in your mind.

The papers Xu sent me were like a brief history of neuroscience. I learned that after Golgi's and Cajal's discoveries, three significant findings in neuroscience—one systematic, one accidental, and one tragic—paved the way for memory research. The underlying theme in all three discoveries is that memories reside in the physical, touchable, damageable material of the brain. These three ideas were foundational to our field of memory research, and to the work that Project X would undertake in memory manipulation.

In the 1900s, the neuropsychologist Karl Lashley became one of the first researchers to systematically search for a memory in the brain. He was fixated on the idea of finding Richard Semon's memory trace, the engram, a term Semon would go on to introduce to the world in 1921. Lashley wanted to know which brain areas housed the electrophysiological bolts of micro-lightning, the biochemical cocktails, the cellular structures that make up a given memory.

In his hunt for the engram, Lashley removed different parts of the rat brain and associated the size of the removed tissue to the intensity of memory impairments. He first trained his rodents to run through mazes to find food, creating a memory of the maze. He would then damage parts of their brains and then place the mice back in the maze to see if their memory was impaired. Damage to a single region of the brain led to slight impairments in navigating the maze, which the animals could overcome; damage to multiple regions of the brain led to larger impairments, but at least parts of the memory still seemed to be intact. Frustrated after a decade of his unsuccessful, albeit highly influential, search for the engram, Lashley concluded: "This series of experiments has yielded a good bit of information about what and where the memory trace is not. It has discovered nothing directly of the real nature of the memory trace. I sometimes feel, in reviewing the evidence of the localization of the memory trace, that the necessary conclusion is that learning is just not possible."

Lashley couldn't find an engram because, as neuroscience would later discover, an engram isn't located in one spot in the brain. A single engram consists of a constellation of neurons *distributed* throughout the brain rather than consisting of one point in a three-dimensional (x, y, z) coordinate system.

Following Lashley's work, in the mid-1900s, Canadian neurosurgeon Wilder Penfield accidently discovered that when it comes to processing memory, not all brain regions are recruited equally. As a neurosurgeon, he would treat several patients with epilepsy, which often originates in areas neighboring the hippocampus, by carefully exposing and

removing the problematic neural tissue. However, to distinguish between healthy and abnormal tissue, Penfield applied tiny jolts of electricity to the brain and reported his patients' responses—some were
innocuous, others were signatures of seizure onset. The patients were
able to be awake (and helpful) throughout the process, since the brain
itself doesn't contain pain receptors; the brain *processes* pain, but having
your brain matter directly touched won't necessarily *produce* pain. After
applying electrical stimulation to areas of cortex near the hippocampus,
some of Penfield's patients reported that they were recalling memories
of past experiences—hinting that *Inception*-like experiments are indeed
possible.

One patient reported that he heard people talking. When Penfield
asked about it, the patient said he could not tell what they were saying
because they seemed to be far away. On the second stimulation, he said,
"Now I hear them. . . . [It's] a little like in a dream . . . like being in a
dance hall, like standing in the doorway—in a gymnasium—like at the
Kenwood High School." He added, "If I wanted to go there, it would be
similar to what I heard just now."

The stimulations sometimes induced extraordinarily detailed experiences. Another patient said he felt like he was "in the dining room—the
front room" and that he could see people "moving about. . . . There were
three of them and my mother was talking to them. She was rushed—in
a hurry."

And another reported: "Some crazy things ran through my mind; I
was younger, at school. I was playing with a polo bat." He said he remembered doing this around age ten.

The memories could be recent too. Another patient said, "My mother
is telling my brother he has got his coat on backwards. I can just hear
them. . . . [This happened] just before I came here."

Penfield's surgeries suggested that electrically activating areas of the
brain near the hippocampus in humans was sufficient to induce the recall of a memory. But what would the *absence* of this particular lump of
brain mean for memory?

Enter Henry Molaison, otherwise known as HM. In the mid-1950s,
just as Penfield was wrapping up his landmark book on the human

brain, this twenty-seven-year-old Connecticut man had been in a bike accident that tragically plagued him with epileptic convulsions. These convulsions became so devastating that he required brain surgery. This single surgery, conducted by William Beecher Scoville, became a cornerstone of memory research. HM had small portions of his brain removed, including parts of his hippocampus and surrounding areas, in an attempt to treat his seizures. But after the surgery, he became suspended in time. Without a functioning hippocampus, he developed an inability to form memories or recall memories from the past, with his amnesia becoming more severe for moments that occurred closer in time to his surgery. Brenda Milner and Suzanne Corkin, two pioneering researchers who studied HM for his entire life, noted that he lost his ability to bridge personal events across large spans of time. But, in turn, he gave scientists a clear road map to navigate how the hippocampus is needed to form memories.

Somewhere between the psychological phenomenon of amnesia and its relationship to the physical deterioration of the hippocampus lay hints of how engrams work.

In 1992, nearly half a century after these three foundational discoveries, Susumu Tonegawa and his colleagues generated mice that lacked a single gene that was shown to be important for neurons to effectively communicate and strengthen their connections—which are key cellular properties for how memories form. Without this gene, the mice couldn't remember how to properly navigate a maze. Soon after, neuroscientist Eric Kandel and his colleagues at Columbia University likewise succeeded at manipulating genes to disrupt entire memories. This is extraordinary, given that a mouse typically has around 35,000 genes, and yet there exist a privileged subset of genes without which some memories break down. These genes, it turns out, are important for plasticity in the brain—the idea that synaptic connections between cells grow stronger and reorganize when processing any bit of experience, such as learning something new or remembering something from the past. And while not every single memory requires the exact kind of plasticity that results from this exact set of genes, the studies here helped neuroscientists explore the cellular rules that the brain follows when storing information for later use.

On the flip side, in the fall of 1999, Princeton neuroscientists Ya-Ping Tang, Joe Tsien, and their colleagues engineered a mouse with a superior memory (compared to standard lab mice) by *adding* more copies of genes that were important for memory. They named their subjects the "Doogie" mice, after the 1990s show *Doogie Howser, M.D.* in which a teenage genius attempts to balance his youthful years with the hardships of practicing medicine. The authors concluded with a bang: "Our results suggest that genetic enhancement of mental and cognitive attributes such as intelligence and memory in mammals is feasible."

In addition to their profound implications, these are noteworthy discoveries because they take our psychological notions of memory and give them a physical framework, beginning with the brain and peering all the way down to cells and the genes they contain. The 1990s thus bridged the world of the molecular with the world of cognition in a massively influential manner that inspired decades of engram research.

It didn't take long, though, for the field to begin applying such genetic engineering to restore health in the brain. Starting in the year 2000, neuroscientists induced or reversed amnesia in rodents with exquisite precision: deleting only a single gene from the mouse brain was enough to dramatically impair memory. Turning that gene back on made the memory deficits go away and, amazingly, making multiple copies of the gene could enhance memory to build smarter rodents. Neuroscientists now possessed the ability to hit the genetic brakes or the accelerator on memory as they pleased. These experiments showed us that a memory could be what Xu called "an unstable Jenga tower": pulling out the wrong block causes the building of memory to collapse, but stabilizing just the right piece can help put it back together. In neuroscience's search for memory, the genes responsible for engrams were demystified with unprecedented detail, one block at a time.

A picture of an engram's molecular components was slowly emerging.

From Plato to Ramón y Cajal and thereafter, science has learned that memories leave a physical imprint in the brain, which binds together all that we have experienced. Memory gives cohesion to our sense of self over time.

Reading through Xu's giant folder of homework, grappling with the significance of how our view of memory has changed over millennia, the fact that the ephemeral act of recollection has a concrete basis was extraordinary to me. I now understood something of monumental importance about the process of remembering: memories are measurable, viewable, and controllable.

Xu and I were about to make use of this knowledge in the lab, using tools that let us control the brain on the same timescale on which it communicates: milliseconds. Like bodies and personalities, neurons come in all shapes and types and rapidly produce all sorts of chemical concoctions. They fire in sub-second idiosyncratic patterns and, as if studying them wasn't already hard enough, in humans there are an extraordinary 86 billion of them. How do we find and target those involved specifically in memory? Could we use a drug to modulate neural activity? The drug would have to work faster than the drugs currently available and would have to affect only the cells holding onto a memory. Was there a new drug that could do this? Drugs are noninvasive, but they're relatively slow and flood the entire brain. Could we use targeted electrical currents? Electricity is fast, but there's no way to deliver it only to cells that harbor a specific memory and not accidently stimulate neighboring cells involved in different memories. Could we try turning genes on and off? This can either be a permanent modification or take several hours to weeks for the effects to emerge. The answer, it turns out, is to use the fastest thing known in existence: light.

But neurons activate or inhibit each other with a combination of chemicals and electricity, so why would *light* somehow activate a given set of brain cells, let alone a memory? A flurry of advances in science and technology over the last twenty years have collectively illuminated the brain's hidden corners with some of the most cutting-edge methods of manipulation. One such technique has not only revolutionized how we probe the brain but our very understanding of how the brain works in the first place. This technique is called optogenetics, and I touched on it briefly in the introduction to this book. Optogenetics was exactly what Xu and I focused on—we could shoot light onto brain cells to activate them and artificially jump-start a memory.

The development of optogenetics relied on the collective expertise of numerous scientists across molecular biology, chemistry, and physics to create a new way of manipulating the brain. These seemingly disparate fields were slowly merging into a field so new that it didn't yet have a name. A scientific revolution was brewing.

The most important contributor to this revolution wasn't any one person, however. It was, of all things, pond slime. For years in the 1980s and 1990s, teams of scientists had been knee-deep in California water while researching an algae's ability to orient itself toward or away from sunlight. There was something special about a specific piece of machinery in the algae that enabled it to sense and respond to light, thereby driving the algae's behavior in one direction or the other.

That piece of machinery was a single protein that today is a household word for nearly every neuroscientist: Channelrhodopsin-2, or ChR2 for short. But the initial studies reporting how ChR2 worked flew somewhat under the radar—they were literally biologically flashy, but perhaps because they dealt with esoteric proteins, and perhaps because they were *just* basic science, it wasn't exactly headline-grabbing news at the time. But these algae contained that one piece of microscopic machinery that, once discovered, would launch a new field of neuroscience, fill chapters of textbooks, and give us unprecedented insight into how the brain works.

Finally, decades of research pivoted on August 14, 2005. The work with algae didn't begin with a neuroscientific application in mind, but one remarkable paper introduced a faster and more targeted method to probe neural activity directly: that method was optogenetics. *Opto*, meaning light; *genetics*, meaning, well, genetics. With this paper, the field began its breathtaking technical and conceptual facelift. Developed in Karl Deisseroth's lab at Stanford, the tools of optogenetics would allow for the control of neural activity in mammalian cells by using pulses of light with millisecond resolution—exactly the tool Xu and I needed to study the brain and memory at its natural speeds.

Xu told me that this paper was *really* important to understand inside and out. One morning in the Tonegawa Lab's coffee room, Xu took to the whiteboard and began explaining how the Deisseroth Lab had engineered brain cells to turn on or off with light by genetically tricking them to install the light-sensitive ChR2. Xu drew an oval-looking brain cell and started adding a bunch of dots to it, which were meant to each be ChR2.

"Think of plants," he said. "Sunlight lands on plants, and they somehow convert this light into energy to stay alive, right? So, by analogy, our laser light lands on ChR2, and ChR2 converts the light into a signal that a brain cell can understand."

He went into more detail: "ChR2 embeds itself on a neuron's surface and contains a pore that can open and close in response to light. Normally, neurons contain their own natural pores, called channels, which open and close to let molecules such as sodium, potassium, and chloride rush in and out and thereby create an electrical charge." He jotted down their elemental shorthand on the board between every breath. Scribbled everywhere were barely legible instances of Na^+ and K^+ and Cl^-.

"This charge builds up as the concentration of ions inside and outside the cell fluctuate until the cell finally reaches a threshold and initiates an action potential, or 'fires.'" This threshold is key and inherent to neurons. Think of it like a neuron's tipping point: when it receives enough inputs and fully charges up, the neuron fires and releases its electrochemical messengers onto neighboring cells.

Xu looked particularly satisfied drawing this process with unnecessarily detailed, three-dimensional lightning bolts. In that moment, watching him, I saw that glimmer of Xu as a kid, talking about science, sketching out diagrams with the artistic detail that he thought would help them stick in my memory.

"The cell releases its chemical cocktails, called neurotransmitters, in a sort of molecular baton handoff to activate or inhibit the next cells in line. Finally, the cell rests momentarily until the next buildup of activity occurs. And so the chain of electrochemical neural communication goes. The trick is that when we embed ChR2 in a neuron, it allows us to artificially control this process."

At the end of his mini-lecture, Xu wrote HOW TO OPTOGENETICS on the board and underlined it repeatedly, before ending with, "It's a masterpiece!" I wasn't sure if he meant the paper or his work of art.

With optogenetics, scientists could finally turn cells on and off while memories were being formed. They could initiate and pause action potentials *as experience begins to leave its lasting mark on the brain*—that is, during the Encoding and Storing processes—all at the same millisecond-timescale on which the brain operates. As the authors note in the paper's final line of prescient text, "Thus, the technology described here may fulfill the long-sought goal of a method for noninvasive, genetically targeted, temporally precise control of neuronal activity." In other words, mind control with light was now in the neuroscience arsenal.

With that landmark paper, neuroscience now had a new-age warp speed technique that felt like it was straight out of *Star Trek*. It kickstarted a paradigm shift in neuroscience, with labs all over the world using optogenetics to unlock many of the brain's previously inaccessible mysteries.

Xu and I could finally illuminate the tangled, dark jungle of the hippocampus and test if we could reawaken a dormant engram . . . with laser beams.

———

By learning about key historical milestones of memory research, I now had a foundation for understanding the engram, including where in the brain to look for one.

The hundred-year quest for the engram began as a prescient idea: experience leaves a physical mark on the brain. First, scientists like Golgi and Ramón y Cajal created the necessary techniques to visualize the brain's constellations of cells. This work paved the way for the neuroscience that followed, by giving researchers access to the brain's mysteries with exquisite cellular resolution. Neuroscientists would learn, for instance, that the brain consists of individual units (cells) that are interconnected through synapses, which themselves physically reorganize (this ability is called neuroplasticity) to convert an experience into a memory.

As the century progressed, studies on both mice and humans began homing in on the areas of the brain that are responsible for memory.

Damaging areas such as the hippocampus produced profound learning and memory deficits; stimulating the hippocampus area generated self-reported stories of remembering. As the field entered the twenty-first century, the search for memory was thriving more than ever before: neuroscientists started to create catalogs of genes and cellular processes that were important for memory, and even invented revolutionary tools to turn brain cells on and off at a sub-second timescale. The tools that are used to study the brain in modern neuroscience, including optogenetics and imaging, developed in lockstep with our maturing questions around the biological nature of memory. How does the symphonic pattern of neuroplastic changes store information? All of this meant that neuroscience could progress toward an era where the mysterious stuff of the mind was now composed of quantifiable connections and patterns of cellular activity.

Xu and I stood at the crest of foundational discoveries that were pushing the field into totally new territory, and now we could advance our understanding of memory further than ever before . . . by controlling one.

2

The Shape-Shifter

FOR AS long as we are alive, the brain never stops working. It absorbs the world around us and creates an internal representation of that world. The brain then adds a layer to these experiences by assigning meaning to them: the brain listens to and loves things that bring it happiness; it learns and remembers those things that form an autobiographical account of being alive; it looks closer at and shakes things that it finds curious.

All of these experiences consist of moments that accrue in the brain over time. From the perspective of the brain, the flow of space and time is as ubiquitous to its own experience as water is to sea life. It is everywhere, always; it never stops influencing. Down to the very forces that govern the universe, our brains move through space and time, and space-time moves through the brain, in a lifelong biological-universal partnership.

Given this complexity, would we even know that we found an engram if it bit us in the hippocampus? It's like asking someone to tell you where life is located in the body—it's *too* general of a question. We have to get a bit more nuanced. Perhaps this complexity is why the manipulation of a memory has taken as long as it has: no matter where we look in the brain, we see echoes of the past, of the brain's space-time travels, etched throughout.

Before getting into the neuroscience of it all, let's briefly put our feet up on our desks and talk philosophy-meets-science, as there are some important experimental boundaries to consider. Like memory itself, the science of memory consists of provisional truths. Neuroscientists look into the brain to measure the physical changes that happen when something is learned and when something is remembered. But this process is complicated by time. Your brain looks slightly different right now than it did a second ago, and it looks even more different than it did a decade ago; similarly, it'll continue to remold itself as the seconds and the years pass. This is what it means for the brain to be *dynamic*. It simply never stops changing. So, if memories in the brain are always changing, how will we be able to find and eventually manipulate them?

Scientists try to make sense of the dynamic nature of the brain, in part, through an organizing nomenclature. Scientists like naming things not just because it's convenient but because it gives us a concrete starting point for understanding that which was once seemingly intangible. For instance, memories of personally experienced events are *episodic* memories and rely on brain areas like the hippocampus; the encoding and storing of episodic memories through learning leads to physical changes in the brain that then make remembering possible. But our memories of childhood *feel* a bit different than the memories we formed last night, yet both exist in the same brain as episodic memories. The former are called *remote* memories because they occurred in the distant past, and the latter are called *recent* memories because they happened closer to the present moment. When you look closely at the cellular processes that underlie remote and recent memories, you'll see important similarities (both recruit some overlapping brain areas) as well as differences (for example, the number of cells recruited in a given area will vary for each).

But it's important to note—like really, *really* important to note—that just because we scientists decided to divide memory into categories, this does not mean that these categories are the ground truth for how memory works. There is no physical line where a remote memory begins and ends and where a recent memory begins and ends. Likewise for the beginning and end of an episodic memory. While it's incredibly useful to have categories that make specific predictions about the qualitative and quantitative nature of recent and remote memories, these categories are simply our best attempt at framing the currently available data. What the data says, as we'll see shortly, is that nearly every single one of our attempts at categorizing memories—episodic, recent, re-mote, emotional, and more—fall short of fully explaining how memory works.

This is because we do not yet have a unifying theory of memory. And that's okay.

I say that the science of memory consists of provisional truths because even though we have powerful experimental frameworks for understanding the brain, we're not quite at a point in neuroscience where we have principles that can safely guide us into, say, a reliable treatment for a memory disorder. What's more, we're certainly not at a point where we can have neuroscience laws that fully explain the mind in biological terms. We *will* discover principles of how the brain works, which will then lead to the deep, physical laws that govern our minds, but for that to happen, we need to explore more than ever before. How memories change over time in the brain is a big—and currently incomplete—part of this understanding.

To give ourselves some perspective, neuroscience is about a hundred years old, so we shouldn't be too tough on our young field. Growing pains are necessary, if not inevitable. By comparison, the field of physics is freaking thousands of years old! Physicists have principles and laws. They discovered them over millennia; they earned their privileged insight into the fabric of reality. We even see such principles and laws in action every day through our mastery over the physical world. We humans do everything from flying airplanes and operating space shut-tles to accidently butt-dialing someone in another country. But when it comes to the field of neuroscience, we've only had a century to build up our knowledge about the brain—imagine where the field of physics was in its first hundred years compared to today. We neuroscientists are more like rookies trying to come up with our version of the Pythago-rean theorem on papyrus. The neuroscientific equivalents of Einstein's relativity and Marie Curie's radioactivity will come much, much later, but they *will* be discovered. As the saying goes, history doesn't repeat itself, but it does rhyme; and progress in neuroscience will continue to evolve just as physics has done, as long as there is something to be ex-plored and someone doing the exploring.

I truly believe that as neuroscience matures beyond its early stages, we'll finally have answers to these questions: What is the fabric of an engram? What are the laws of memory?

While we're only beginning to understand how the brain changes over a lifetime, one of the milestones in the quest to manipulate a memory has been finding where in the brain a memory is located at *specific moments*, even if its cellular identity is constantly shape-shifting.

This chapter is about the nomadic nature of memory. Neuroscience in the twenty-first century has been able to successfully take snapshots of memories as they age in the brain, from memories formed in early adolescence to those formed late in life. We're even able to turn these different memories on and off. Despite their constant cellular travels, we are now learning that memories move around in a predictable pattern in the brain, especially as they age, which brings us closer in our quest to fully controlling their activity.

For now, let's inch our way forward, one memory at a time. Let's look in this chapter at how different types of memories are represented in the brain and, specifically, how they move throughout the brain over time. I begin by taking a look back at a key memory of my own that opened this book.

———

May 26, 2011.

Xu and I went into our lab's vivarium to gather five mice. Back in the lab, we placed each mouse, in turn, in our almond-scented box, where we were hoping to witness the artificial activation of a memory. But when we shined the blue light into the hippocampus of the first mouse, nothing happened. When we did the same to the second mouse, it didn't even flinch. The third mouse: zero. The fourth mouse: nada.

We connected the laser to shine blue light into the brain of the last mouse. Everything now depended on mouse OF5—so named because it was the fifth rodent we had implanted with an optic fiber. *Click.* It immediately froze when we shot light into its hippocampus. And Xu and I froze too.

Embedded within a tiny corner of the brain are memory-bearing hippocampus cells that are enough to trigger memory recall. Memory manipulation was no longer the province of science fiction and Hollywood anymore on that day—it was officially in our toolbox for understanding how the brain operates. The lab playfully began calling Xu and me *Team X* because we had gotten the first bits of evidence that Project X was the real deal.

It's breathtaking when science works.

Now we really had to start playing detective. Why did mouse OF5 freeze when we delivered light into its brain but OF1 through OF4 didn't? What would we do next, knowing that we just saw our first real-life demonstration of the artificial activation of a memory? Team X got back to work.

"We need to compare the brains of each of the five animals to see what made the fifth one different from the rest," he said.

When we looked under the microscope, we could see ChR2 fluorescing green in the cells of the hippocampus in the first four mice, OF1 through OF4. The ChR2 appeared in exactly the right region that we were aiming for, but these mice didn't show evidence of light-induced memory recall. Within the brain of mouse OF5, however, we encountered a serendipitous surprise. In OF5 we accidently had botched the surgery and had injected ChR2 into an unintended part of the hippocampus, slightly deeper in the brain than we meant to. The space between these two different parts of the hippocampus was half a millimeter, about the thickness of five sheets of copy paper stacked on top of each other, but it is in that microscopic space that our discovery on artificially manipulating memories originated.[1]

The green glow from an image we took of ChR2 in the hippocampus of mouse OF5 lit up the computer screen; I stared at it, while Xu hunched over the attached microscope and looked through its eyepieces to study the brain slice more closely, the reflection from the green fluorescence occasionally refracting through the lenses of his glasses. He perhaps let the feeling sink in that a real, physical memory was looking back at him for the first time. I'd like to think that Xu was thinking the same thing I was: *At some point in life, we all get our fifth mouse.*

The two of us mapped out a schedule for the next few months and split the work right down the middle. We may have successfully reactivated a memory in mouse OF5, but now we needed to replicate our results. I booked as many time slots in the surgery suite as I could.

Since we had a rule in the lab that the first one in the surgery suite got to control the radio for as long as they were in the room, I was the annoying lab member who got in by seven in the morning to crank up my favorite local music station instead of the audiobook or morning talk show my lab mates always chose—these never failed to put me half to sleep. Whether they liked it or not (narrator: they did not like it), the Tonegawa lab learned more about Top 40 music in those months than they ever cared to do again.

Our experiments kept working, and we kept pushing the project forward. On Thursdays, we went to the Thirsty Ear Pub a few blocks from our lab for karaoke night. Xu would say, "I'd rather do surgeries for twenty-four hours a day for the rest of my life than get on stage," but this didn't stop me from taking the opportunity to butcher "Party Rock Anthem" every chance I got. This late-night lifestyle became a pressure relief valve for the stress that we endured during the day.

I remember the exhilaration I felt when, one Friday, Xu and I finally presented our results during the Tonegawa Lab meeting. These lab meetings were a weekly opportunity to do science out loud: to provide encouragement in real-time, to troubleshoot problems as a lab, and to evaluate each other's work to make sure our projects included the proper experiments. They were also intense. "Extraordinary claims require extraordinary evidence," Carl Sagan pointed out, and Susumu, along with the lab, expected both every Friday.

I wasn't initially excited about presenting in front of the entire lab. It didn't help that I had a gut-punching fear of public speaking. While preparing our presentation, Xu agreed to interject if I ever looked like I was about to have a panic attack. He also told me our goal was to tell a clear story the lab could understand. He said he used to practice his talks by pretending he was explaining his research to his family in China, and then he said something to me in Mandarin.

"Was that a joke or an insult?" I asked, laughing.

"I said: 'I study the brain and want to reactivate a memory.' Now say it to me in Spanish."

"Uhh . . . *estudio el cerebro y quiero reactivar una memoria.*"

"Perfect. I think. Well, I don't know Spanish. Anyway, now tell me how you're going to reactivate a memory in the brain."

"*Quero estudiar el cerebro y. . . . para reactivar una memoria tengo que . . . uhh. . . . El cerebro tiene células . . . y. . . .*"

A linguistic *gato* had gotten my *lengua*. Xu's lesson was something that would change how I delivered presentations. I'm fortunate to have Spanish and English as my first languages, so Xu told me that when I worried that my presentation was going down a confusing jargon-filled rabbit hole, I had to explain that slide or idea to myself in Spanish. Make being bilingual a superpower, he said. The exercise forced me to erase all the technical words and reinterpret them with metaphor.

I had no idea how to say *Channelrhodopsin-2* in Spanish. Or *engram*. Or *hippopotamus*, let alone *hippocampus*. But I did understand *what* I was trying to say, and I was trying too hard by using terms that sometimes confused me and, by extension, doubly confused an audience. I learned to first interpret an idea in English and, if it felt too loaded with technical terms, I would translate it into Spanish and put it on a verbal diet. Then, I would reverse-translate it from Spanish back to English and—*¡qué magnífico!*—I had a new way of describing what was once incomprehensible. In true Aristotelian form, Xu was the walking embodiment of the saying, "Those that know, do. Those that understand, teach." For instance, which one of these sounds more intelligible?

Xu and I injected an inducible and activity-dependent c-Fos-driven Channelrhodopsin-2 construct into the dentate gyrus of the hippocampus and optogenetically reactivated these corresponding cells with 473nm light at 20hz with 15ms pulse widths over 180 seconds to produce freezing behaviors.

(Filters all of this through Spanish.)

We artificially reactivated cells in the brain that held onto a memory.

When Xu was done with his half of the presentation, introducing the idea of manipulating an engram, I took over and presented a single graph that summarized the batch of experiments we had done since March with mouse OF5. In mice OF6 to OF10, when we delivered optogenetic stimulation into the hippocampus, the animals also froze just like OF5, and when we turned off the light, their behavior went back to normal. When we optogenetically stimulated these cells again, the animals froze, and when we flicked the light off, the behavior once again went back to normal.

"I think we have a way to literally switch memories on and off in the mammalian brain," I said, my voice shaking a bit. I had no idea if the lab members would believe us.

"WHY DIDN'T YOU START WITH THIS!?" one of the senior postdocs shouted.

The other lab members encouraged us to submit the paper in time for an upcoming beast of an event—the Society for Neuroscience meeting in Washington, DC, that would be packed with more than 30,000 people in our field.

At the end of lab meeting, Susumu asked us to think about where we wanted to submit the paper, which was his way of saying, *Pick the best damn journal you can.* He *believed* our extraordinary claims based off the data we had presented.

Xu looked over and said to me, "Five more years of lab meetings to go!"

———

As my body grows older, my wrinkles become more apparent (I don't want to talk about it), and my beard starts to become speckled with white hairs (I *really* don't want to talk about it). Do my memories have similar physical features about them that change in the brain as the inexorable force that is time moves forward?

Let's travel back to memories I made when I was six years old to find out. When I was in kindergarten, my parents would take my older brother, my older sister, and me food shopping at the local Market Basket on Friday evenings. They had a hundred dollars to feed a family of five for the week, and somehow my parents made it work. This, however, didn't stop me from yanking at my mom's sleeves and putting on my best wide-eyed gaze and plumped-face frown in an attempt to convince her to buy me a *Toy Story* coloring book or a *Mighty Morphin Power Rangers* action figure. These memories have existed in my brain for decades, aging with me as the years go by. I would find out much later, when I was in college, that the weeks when this worked were the ones when my dad had worked enough hours at his third job to make extra money to spoil the family. The other weeks, there just wasn't enough money for extras beyond college savings, which my parents considered as important as putting food on the table.

Yet all the details that I stored spring back to life each time I close my eyes and think all the way back to my childhood or back to my friendship with Xu: my dad riding the shopping carts with me up and down the Market Basket aisles, and Xu and I learning how to row along the Charles River together; my brother and sister scheming away on how to throw a house party when my parents went to El Salvador to visit family, and Xu and I going away to New Hampshire on a ski trip when Susumu was out of town; my mom bribing me with movie tickets for every *A* I got in school at the beginning of each semester, and Xu and I going to the midnight premiere of *The Dark Knight Rises*. These memories are intertwined at the deepest, biological level because, somehow, I can mentally time-travel back to my parents and then fast forward through the strands of neural space and time to Xu, and back again.

This is because the brain has a way of physically connecting emotionally poignant memories. To watch this process in action, neuroscientists at a memory research lab at the University of California, Los Angeles, first had to take videos of memories in the mouse brain using miniaturized microscopes. (The devices are so small they can balance on your pinky and contain a lens that's less than a millimeter thick, rivaling the thickness of, say, a credit card.) These scopes are the very technology that neuroscience needed to track individual memories in the brain. To announce such technology, in 2011 a Stanford team published a paper that proclaimed that their microscope offers a transformative piece of technology that can be used across species and even enable portable diagnostics and large-scale screens of the brain performing almost any task and thinking any thought. Once these kinds of scopes are surgically implanted, they provide a window into the brain's mysteries unfolding in real time.

It's quite beautiful: the videos these scopes provide look like glimmering stars reflecting on the undulating surface of an ocean. Only a few years later in 2013, the Stanford team successfully visualized how thousands of brain cells behaved over the course of weeks when animals experienced the same memory repeatedly, which was a virtually impossible task without their new miniaturized microscopes. They discovered that a small percentage of hippocampus cells were always active for the same memory even if weeks had passed since it initially formed. Then in 2016, the UCLA team led by Denise Cai pioneered the development and use of open-source miniaturized microscopes, or miniscopes, and discovered that when memories were linked, they did so by existing in the *same* brain cells, which provides us with biological evidence of memories physically living together in shared neuronal real estate. I find this profound. It follows that, despite having never met, my parents and Xu interact in a very real sense in my brain by living in close cellular proximity.

The memories I have of Xu and my parents also have a common thematic denominator, rich with emotional significance: that of family. Think back to my recounting the memory of visiting El Salvador when I was a young boy, for example. Memories involve the rich details—the personal episodes—of our past, stretching from when you end this sentence all the way back to your very first memory, which can be intertwined with the feelings—the *emotional* components—of our past, including family trips abroad as well as trips to the grocery store. If memory is like a full-color flip-book, then the episodes we experience are its sketches, and our emotions are the vibrant colors of those sketches. Our recollections are our brain's way of rapidly flipping through each page of the book and animating a once-lived moment.

The emotions and episodes that comprise our memories recruit teams of cells, or *ensembles*, distributed throughout brain areas that communicate with each other and form *circuits*. Remember Henry Molaison, aka HM, from chapter 1? His experience taught us that the hippocampus contains key clusters of cells needed for episodic memory.

Damage to the hippocampus compromises our ability to remember episodic memories, but not entirely. HM couldn't remember events that he experienced after his surgery, and yet many memories that he formed prior to his surgery remained intact, such as his childhood, his wife's name, the emotions surrounding each, and how to ride a bike. How is this so? How is it that some memories seemed to rely on the hippocampus, and yet others seemed to live elsewhere in the brain? The brain, it turns out, contains *memory systems*, meaning that we have different types of memories that together make the richness of experience possible. Just how many of these systems we have is a topic of intense debate, but we at least know that memories of, say, a partner's name, the emotions we feel, the episodes of our lives, and riding a bike, each recruit a complex array of circuits in the brain that make them possible. Memories like our partner's name are *semantic*, meaning that they are fact-based knowledge about the world; memories of our first-person experience such as rocking in a hammock in El Salvador are *episodic* in nature; how any of these make us feel are aptly termed *emotional* memories; and finally, the physical act of riding a bike relies on *motor memory*. All manifest distinctly across the brain and are probably part of a larger function of the brain—namely, to make sense of our ongoing reality in the service of living into the next moment.

The concept of memory systems provides researchers with useful distinctions about the kinds of memories the brain can make. For example, HM's memories of his childhood were intact, despite no longer having much of a hippocampus. But presumably, when HM stored those memories as a child, they would have engaged different brain systems than they did when he recalled them as an adult. Will my memories of Xu and my family shuffle around my brain as they get older? The idea that memories transform over time—and engage in different circuits of the brain as they do so—has a long and remarkable history. Cases like HM suggest that, while newer memories depend on the hippocampus, older memories are spared by such damage and therefore must depend on a brain area that is not the hippocampus. That narrows down the search for old memories to *oh, just literally the rest of the brain.*

Let's start with a period of time that a student recently called "the late 1900s," much to my embarrassment. In 1999 in Robert Jaffard's lab at the University of Bordeaux, neuroscientists Bruno Bontempi, Catherine Laurent-Demir, and Claude Destrade published a paper in *Nature* titled "Time-dependent reorganization of brain circuitry underlying long-term memory storage." Like humans, mice have memories that can last a lifetime. The team first trained their mice to remember where food was located in a maze and then tested their memory 5 days later (a short amount of time after learning) or 25 days later (a long amount of time after learning). These experiments tested whether or not different brain areas were recruited in mice recalling the recent 5-day-old memory compared to mice recalling the older, or "remote," 25-day-old memory.

The approach they used can be described two different ways. First, in their own words: "(14C)2-deoxyglucose was injected 1 min before the beginning of either the last training session or the retention sessions. . . . Autoradiographs were analysed by semi-quantitative densitometry using a BIOCOM 200 computerized image-processing system."

Or . . . in English: they made active cells glow. More specifically, they made it so that cells that were active during the recall of a recent or remote memory now gave off a detectable signal that could be seen by an experimenter.

The results were stunning: in the recent memory group of mice (that is, the 5-day-old memory group), the hippocampus was highly active; in the remote memory group of mice (the 25-day-old memory group), the outer mantle of the brain was active, namely, the cortex. It's as if, somehow, the hippocampus acted as a temporary repository for encoding recent episodic memories and, over time, the cortex became a permanent storage site for older memories.

That means the brain's mental time machine needs to visit the cortex for a remote memory or park itself in the hippocampus to visit a recent memory. As time passes, as a memory begins to move beyond its initial encoding and storing phases, various circuits in the brain strengthen their communication and become responsible for transforming the memory into something permanent. These memories then exist across multiple circuits in the brain, and these circuits collectively form what neuroscientists call *systems*. Cells make up ensembles; ensembles make up circuits; circuits make up systems. When memories become stable, when they engage various systems and solidify their presence in the brain, they have become *consolidated*. They become resistant to hippocampus damage over time as they move throughout the brain. It's the reason why Drew Barrymore's character in *50 First Dates* still remembers who she is but can no longer remember the events just prior to her car accident. The same logic applies to Leonard Shelby in *Memento*: he's unable to remember events that occurred just minutes ago, but he still somehow remembers the names and faces of his friends and family.

One of the most influential theories in memory neuroscience is fittingly termed *systems consolidation*—the process by which an engram is temporarily encoded by and stored in hippocampal circuits and then is stabilized and transferred to non-hippocampal circuits, such as the cortex. In that same 1999 issue of *Nature*, Edmond Teng and Larry R. Squire at the University of California, San Diego, published a study of humans who had damage to both hippocampi and observed the same thing the Jaffard lab team saw in rodents: "We tested a profoundly amnesic patient (E.P.), who has virtually complete bilateral damage to the hippocampus and extensive damage to adjacent structures. . . . We asked him to recall the spatial layout of the region where he grew up, from which he moved away more than 50 years ago. E.P. performed as well as or better than age-matched control subjects who grew up in the same region and also moved away. In contrast, E.P. has no knowledge of his current neighbourhood, to which he moved after he became amnesic."

Their results showed that recently formed memories in humans relied on the hippocampus, while older memories were stored elsewhere. So even though Xu and I were able to reactivate a memory that was only a couple of days old by stimulating the hippocampus, we had no idea if these experiments would work on memories of different ages or complexity as they transform themselves throughout the cortex.

"If you want dogma, read up on systems consolidation," Xu advised me as he prepared to discuss a recently published paper with the Tonegawa Lab. As part of our never-ending quest to catch up with all the discoveries in memory research, Xu and I would partake in the lab's monthly journal clubs, when one lab member would present a newly published paper for the rest of the group to discuss. Xu chose a recent paper and valiantly called dogma into question.

"Everyone in our field has an opinion about systems consolidation, and we probably aren't even close to figuring it out," he said, amused. "I think it's a case of researchers wanting something to be true for so long that they just begin to believe that it's true . . . even though the brain is a bit more complicated than we've been led to believe. If we don't need the hippocampus to remember really old memories, then why does the hippocampus still light up like crazy when a person is getting a brain scan while remembering an old memory?"

Xu was referring to the seemingly paradoxical findings that showed that patients with hippocampus damage (like HM or E.P.) couldn't encode new memories or recall new events but *could* remember much older memories; however, when human subjects with intact hippocampi got their brains scanned and were asked to recall old memories in vivid detail, large increases in hippocampus activity were detected.

In one of his slides, Xu showed two pictures, each containing a brain scan of an intact human hippocampus: one was of a person recalling a recent memory, and the other was of a person recalling a very old memory. They looked practically identical. So why is the hippocampus active when encoding and remembering memories of all ages, yet damage to the hippocampus only impairs *newer* memories?

Xu's presentation reminded us of the current state of memory research: by the turn of the twenty-first century, the dogma of systems consolidation was that a hippocampus was needed to encode and store new memories and that as these memories aged in the brain they were shuffled off to the cortex. But optogenetics gave us a completely different tool to turn the hippocampus off with unprecedented precision and to test for its role in processing memory.

Neuroscientists Inbal Goshen, Karl Deisseroth, and their team set out to use optogenetics to directly test the time points at which the hippocampus was needed to recall a memory, recent or old. As we've seen, the traditional view, which the team replicated, was that damaging the hippocampus with a lesion or a drug should only impair memories that were recently formed. The authors then used a tool of optogenetics that one can think of as "ChR2's other half," as Xu called it. Discovered in 2007, NpHR (halorhodopsin from *Natronomonas pharaonic*) responds to light to *turn brain cells off*. ChR2 is a way to flick a switch on in the brain; NpHR is a way to flick that switch back off.

Previous experiments used drugs to inhibit brain cells, a process that takes several hours, or lesions, which permanently damage the brain. Goshen and the team went on to ask: What if we used the rapid speed of optogenetics to inhibit the hippocampus for just a few seconds? When they inhibited the hippocampus with optogenetics weeks after a memory had been formed—at the time point where the memory had now supposedly become the province of the cortex—the swift inhibition surprisingly *impaired* the old memory. Rapidly inhibiting the hippocampus revealed that it was in fact needed to recall an old memory. But because drugs and lesions are much slower ways of inhibiting the brain, they give the cortex enough time to "compensate" for an impaired hippocampus.

Goshen's results meant that inhibiting a brain region for a long period of time with a drug or lesion was dramatically different from inhibiting it for a very short period of time optogenetically. As Xu proclaimed at the end of journal club: "This study is groundbreaking because it shows us that even the tools we use to study the brain can bias how we proceed to understand the brain itself. It's as if we used to amputate parts of the brain to see how it would affect memory, but now we can take a scalpel and finesse how we dissect memories of all ages." He was right: in the years that followed Goshen's study, several labs went on to show that the subsets of cells within the cortex and hippocampus engage in a real-time dialogue with each other at the time of learning and that their activity is continuously refined as a recent memory matures into a more remote one. This fine-tuning of communication between cortical and hippocampal cells is a multipurpose process: the hippocampus is not only engaged by recalling the remote past, but also the very cells involved in recalling these older memories enable the brain to incorporate new information into the memory itself. At any given moment, ensembles of cells throughout the brain, including in the cortex and the hippocampus, are sampling our ongoing experience to extract new information and update our brain's representation of the past, no matter how near or far it may be. I couldn't help but wonder what the implications were for my brain. I still wonder about that. Does it mean that my childhood memories of my parents live in my cortex, and when I recall them they're temporarily moved back to the hippocampus to be enriched with episodic liveliness? As I grow older, is Xu slowly being shuffled from my hippocampus to my cortex and, if so, will he reside next to my parents? Or have these memories already found each other in the brain, tied together by a magical mnemonic bow?

I think I have an answer, and it comes from the most biased, least controlled, and subjective perspective of all—my own—but rest assured it's at least supplemented by one of the most badass composers of our time: Hans Zimmer. I remember that when Xu and I were writing our first memory reactivation project together, the movie *Inception* had recently come out, and I'd listen to the soundtrack on repeat while working in the lab. The final song of the soundtrack and of the film is called "Time," and it holds the answer to just how close in my brain Xu and my parents really are. All I have to do is hit "Play" and. . . .

———

I'm slouched over my lab desk working on version 60 of our memory reactivation manuscript, while Xu is a few lab benches over polishing up the graphs that will go into the paper. There's the hoppy aroma of Sam Adams Boston Lager on my breath. A lab freezer hums to my right, rush hour traffic is at a standstill outside the windows to my left, and I can feel my feet kicking aside a pair of running shoes and a baseball glove I had left underneath my desk. "Time" plays on repeat. The running shoes were a gift from my parents who knew that I was training to run in the Boston Marathon. My dad gave me the baseball glove for my thirteenth birthday, and I've held onto it ever since. The running shoes remind me of a time when I'd play hide-and-seek with my mom and run as fast as I could as she counted to 100; the baseball glove reminds me of a time I mistakenly threw a ball past my dad's reach and broke the first-floor window of our house.

I sit back in my chair and stare at the monitor, while more of these memories take hold of my focus. The glove sparks memories of a softball game when Xu himself threw the ball past first base and into the stands (which, thankfully, were empty because this was division-negative-infinity MIT softball that we were playing, after all). He asked me to stay after the game to practice fielding ground balls and proceeded to throw god knows how many of them to first base for an hour. My dad was also persistent in his pursuit of perfection: as part of his lifelong routine before his soccer games, he'd spend an hour practicing corner kicks and penalty kicks. He imparted to me the importance of a disciplined approach to learning. Ever since I was six years old, I had to practice piano every day for an hour. I did this for over a decade while living at home. Xu applied a similar philosophy: do it again and again until it's right. But he trusted *me* to coach him through each ground ball and to repeat the process until it became second nature. "Learning is a forever thing," I told Xu, and it's the same bit of advice my dad would tell me before I played the piano for the family. The first song I learned to play from the *Inception* soundtrack was "Time," which I played for my parents at a Christmas gathering. And it plays on repeat while back at my desk, working on version 60 of our manuscript.

———

"Time" is the bow that ties together my memories of Xu and of my parents. It's subjective evidence that I can't think of one without the other, meaning that in a very real sense they *must* be biologically linked in my brain. When Xu came into my life in 2010, my brain quickly assimilated him as a meaningful member of what it considers my family, perhaps even funneling my neural representation of him in some of the same cells that hold onto my memories of my parents. I'd posit that the emotions I share for both are also shared at the cellular level. There are cells in my brain, rich in their ability to reanimate the past, with Xu and my parents at their center, with "Time" as an ever-present backdrop of music that synchronizes each. It lets me time-travel back to a moment and to wade through any related memories I happen to have, no matter how deep or wide they are.

Given that memories shape-shift in the brain as time and experience pass, is it possible to know with exact precision both the whereabouts and contents of a memory at any given moment in the brain? In physics, Heisenberg's uncertainty principle states that there are very real limits on what we can simultaneously physically measure: we cannot know both the position and the velocity of a particle with exact precision, for example, if both are measured at the exact same time. The more we know about one, then the more uncertain our measurements of the other become. Likewise, and by analogy, is there an inherent uncertainty to what we can know about the biological position and movements of a memory as it pinballs throughout the brain over time? If so, this would mean that the more we know about the cellular properties of a given memory at a single point in time, then the less we'll know about its precise contents—we can measure cellular activity and gene expression at a moment's notice all we want, but can we simultaneously measure the exact phenomenology of memory? Similarly, the more we know about the rich, subjective contents of a given memory, then the less we'd know about its molecular composition. We can self-report episodic details of a trip to El Salvador with as much vivid detail as we'd like, but can we at the same time decode the molecular biology of the corresponding memory?

This is where physics and neuroscience may conceptually part ways for the time being because I truly believe that we can have our Heisenberg cake and eat it too. To answer these types of questions, we'd need to measure multiple cellular snapshots of a memory as it changes over time and tether each of these snapshots to their corresponding psychological readouts. In other words, we need biology and psychology to continue to come together as the unified science of the brain that neuroscience offers.

The implications here are transformative: if we peer back into the cellular snapshots of experience, I believe we can excavate even more from a moment than we ever thought possible, and even relive all of what has been forgotten. I don't remember what I did before or after working on version 60 of our paper, but that doesn't mean these experiences aren't in the brain's lost-and-not-yet-found section. Maybe there's another piece of music out there that is braided with these moments, and my brain is just waiting for the right trigger to access them. There may indeed be a way to extend how much of a memory we remember by accessing all of its physical properties.

The work that Xu and I did made it possible to retrieve specific moments of the past with the flip of a switch. But we wanted more—not just to control one experience but to control *all* of experience, to bring more memories back and fill in all their glorious detail. For me, beneath the science of it all, there was a personal pull of nostalgia, of longing for a good memory to be brought back with as much sensory and emotional richness as I had originally experienced. At this point in my PhD, I had no hesitation committing to a future where I could again feel the cold pebbles under my feet in the lake behind my grandparents' house or stargaze with my parents in a village without any light pollution or hold onto the back of a metal grocery cart at Market Basket while my dad pushed. A world where savoring the moment extends into savoring it alongside the past is a world worth creating, I thought. And with *Inception* on our minds, Xu and I began questioning whether we could implant memories or whether we could gain access to some hidden corner of our mind that had seemingly forgotten an episode of the past. Just how deep into a memory could we go?

3

Do You Want a Spotless Mind?

WE SHAPE our understanding of the world by experiencing it—through moments as simple as dining out, going to class, scrolling social media, or taking a familiar route to work. Embedded within these experiences are memories layered with feelings, coloring how we perceive and remember the world. Xu was born on November 23, 1977, but when I think of this fact, I immediately recall that he enjoyed seafood, felt most at home in biology classes, found social media unappealing, and took the Red Line to Kendall Square each morning to get to lab. These details form an emotional tapestry, each memory bound to a feeling. In a way, then, we can think of memory as the sum of emotional impressions that shape us.

Our memories, therefore, are more than just recorded facts; they carry an emotional weight that imbues them with feeling. Meeting Xu by his lab bench in 2010 or remembering an evening with him at the Prudential in Boston—both of these events exist in my brain tied to particular emotions they sparked. While the details may remain static, our emotional responses are more fluid. Joy, nostalgia, even sadness can transform these memories and how we experience them. So our brain is not merely preserving moments; it's preserving how they made us feel, allowing emotions to become core components of what we remember. How do emotions, then, alter or intensify the memories tied to these experiences?

This chapter is about how memory is more than just a record; it's the emotional resonance embedded within our thoughts that pulls us and pushes us—a personal anthology in which feelings like happiness, pride, or regret mark specific experiences, adding significance beyond the details of what happened. Altering memories, then, is about more than just altering facts. It involves altering emotions, changing that wonderful tapestry that gives our lives meaning.

We've explored some of the categories that scientists use to label certain kinds of memories—especially episodic memories, which consist of our personally experienced events. Let's go to the textbooks and unpack this a bit more.

In 1972, psychologist Endel Tulving put forward a theory on memory storage, suggesting that there's a distinction between the types of memories we hold onto, including an *episodic* category that consists of the memories of our personal experiences. He asks us to consider the possibility that there are fundamental psychological and biological differences between different types of memories in the brain and that we should consider each type of memory separately until we reach a point where we can understand how they interact.

Episodic memory here includes an event ("what happened?"), its location ("where did it happen?"), and a time frame ("when did it happen?"), so that the brain can make sense of *why* anything happens. Emotional memory consists of the memory of an event and the feelings it evokes. Think of it this way: episodic memory is our first-person memoir of the world, and within this memoir are events that our biology considers important or salient enough to trigger an emotional response. Researchers believe that emotions can originate from both our body and our brain. They can be triggered by our bodily responses to stimuli in the world, such as increases in heart rate and sweating that our brain then tries to make sense of, or by our brain's response to stimuli in the world, like when the brain detects a threat and then signals to the body to release stress hormones, which go on to change our behavior accordingly. Emotions, in other words, can be built in the body and rerouted to the brain for us to perceive, and they can be built in the brain and rerouted to the body to alter what we're doing. They're a brain-and-body-wide phenomenon, which only underscores their biological importance. But when we recall an episodic memory and the emotion associated with it, we often don't perceive whether the emotion began with the brain or the body, and we often don't separate the facts of an episode from the emotion attached to it.

For instance, let's imagine you're on *Jeopardy* all of a sudden. You've made it to Final Jeopardy and all your hard-won money is on the line. You and your opponents are given the following answer, for which you must come up with the question for the win: "Neuroscientists use this term to refer to the physical manifestation of memory in the brain."

YOUR OPPONENT: What is *enchilada?*
HOST: No, sorry, but yum!
YOU: What is *engram?*
HOST: Correct!

You're elated now! The physical footprint of this engram in your brain is packed with emotion; the story we tell about any achievement (or failure) is infused with how it felt at the time. If you won a lot of money on *Jeopardy*, then presumably your story would contain an obvious layer of excitement. Alternatively, if you put up a big stinker of a performance and left empty-handed, your story would likely feel like a wet blanket. Our ability to feel emotions, not surprisingly, connects us to an experience on a much deeper level. When we remember, we feel.

But emotions do not just imbue our lives with color and richness. Emotional memory impacts the decisions that we make in the present. Emotions may even guide our choices more than the events themselves ("It just *felt* like the right thing to do"). Some argue that it's the main driving force behind nearly every decision we make, regardless of whether or not we're aware of their source or their presence. It's powerful stuff: emotions aren't just a reaction; they're intertwined with memory itself, shaping our physical responses, influencing behavior, and affecting our relationships. They change everything from our body's physiology, to how we behave in the world, to the very relationships we forge and lose.

As Tulving intuited, all categories of memories influence each other in the brain. Episodes and emotions neuronally tug and pull at each other; they can make each other work better or interfere with each other; and they all make measurable, physically detectable changes in the brain. When their corresponding cellular activity goes quiet, the memories go away—they play neuroscience's version of hide-and-seek. When this activity returns under the right circumstances, the memory reveals itself, changing the brain and body in sometimes unexpected ways, as I explore in the pages that follow.

The conceptual framework of this chapter is that episodic memory is about experiencing and that emotional memory is about the (positive and negative) significance of it all. Contemporary neuroscience has been able to identify the cellular ensembles responsible for each type of memory and to study how they change over time. That the brain braids these kinds of memories together so seamlessly is perhaps why any given moment in life can be as magical or miserable as it wants, but one thing is for sure: the stamp they leave on the brain is as memorable as it is manipulatable. So, what happens when neuroscientists successfully disentangle an emotion from an episode and then manipulate each to change their contents?

Our journey into memory manipulation brings us now to how the brain feels, which begins by understanding memory as an emotional archive—our personal history, shaped not only by what happened but by how we felt when it happened. I'll deconstruct the phenomenon of re-membering by showcasing the varieties of memories we all have: memo-ries of our personally experienced events and of the emotions that they come with. Our tools are so good that we can erase entire memories from the brain—events and emotions—and produce a spotless mind. On the flip side, we can also activate different kinds of memories, even ones that were thought to be lost forever, including their long-buried emotions. Finally, I'll dig deep into some of my own difficult memories to pose a couple of thorny questions to us all: If we had the power, would we really want to erase entire moments of our past? Are there certain moments we'd want to reawaken out of dormancy to relive?

———

Some papers in science have titles that let you know they're going to be instant classics, and Xu and I shared a common favorite that almost immediately became a semantic memory the second we read its title. Its title is simple, it states exactly what was accomplished, and it sounds like it's from a sci-fi movie: "Selective erasure of a fear memory." Take a moment to incorporate this semantic behemoth of a discovery into your mind . . . and now we can let all the necessary emotions come flooding through.

The paper has its analog in Hollywood. In *Eternal Sunshine of the Spotless Mind*, Joel and Clementine are former lovers whose previous relationship fell apart and caused them irrevocable emotional damage. When Joel finds out that Clementine has hired Lacuna, Inc. to erase her memories of him, he decides that the only way to prevent a lifetime of grief is to have the company perform the procedure on him as well. The movie unfolds as his memories of Clementine are erased, one by one, and we witness a raw irony: sometimes removing that which hurt us the most means losing a part of our identity.

Eternal Sunshine poses a double-edged, sci-fi scenario, which was becoming more of a reality to Xu and me as our research progressed: the inevitable consequence of erasing a memory is removing a layer of experience that sculpted us into who we are today. But maybe Hollywood threw the baby out with the bathwater when it came to memory erasure. Memory is the brain's retelling of the world, so what if we could remove only certain components of a memory, such as its semantic details or episodic dimensions or emotional hues, without having to get rid of the entire thing? Joel and Clementine were getting rid of their entire memories, or engrams, of each other. But what Lacuna needed was a procedure that could target only specific aspects of a memory, such as the heartache that comes with a messy breakup, as opposed to deleting the entire intricate web of experience that two people share. Biology is in our favor here since the varieties of memories that we have come with unique biological features. For instance, episodic and semantic memory are made possible by different patterns of neural activity stretching across unique territories of the brain; emotional memory engages the vastness of the brain in a unique pattern as well. Even though there is no exact line in the sands of the brain where episodic, emotional, and semantic memories begin and end, as we saw in the opening of chapter 2, we do know that there are enough measurable differences between each that neuroscientists have successfully used these categories to target one kind of memory while leaving others relatively intact.

By deconstructing memory into its more nuanced pieces, the possibility of deleting only some parts and not others becomes more of a reality. That real-life selective erasure of a fear memory from the brain occurred for the first time in Sheena Josselyn's lab at the University of Toronto in 2009. Josselyn's work is the neuroscience equivalent of the yellow line in Olympic swimming that lets viewers know the position of the world record, of something remarkable that is just within reach. Josselyn, along with her lab members Jin-Hee Han, Steven Kushner, and colleagues, published a trailblazing paper reporting that they could target the brain cells processing a fear memory, kill only those cells, and successfully erase the emotional memory itself.

The experiment is ingenious: first, they genetically "biased" a small fraction of brain cells in an almond-shaped area called the amygdala to become highly active. Based on decades of prior research, the lab had substantial reasons to believe that the amygdala was one of the brain's key emotional loci—it becomes particularly active in response to all sorts of salient scenarios, such as experiencing fearful stimuli or a bout of anxiety, and its absence can lead to a sense of "fearlessness."[1] The Josselyn lab introduced extra copies of a gene into a fraction of amygdala cells that would make them much more likely to be "on" or "firing" while a memory is being made. Because those cells were more likely to be active compared to neighboring, nongenetically modified cells, they acted like neuronal sponges in the brain—that is, they were the first to readily absorb a memory as it was being formed. This meant that, even before forming a memory, the authors could already pinpoint which cells would become a part of the memory. They could now artificially "allocate" an engram to a predetermined set of cells. The first step to erasing a memory was therefore funneling a memory, as it's being formed, into a predetermined group of cells in the brain.

Josselyn and her team then introduced another set of genetic instructions: they made it so that the highly active brain cells now contained a molecule which, when activated, caused a chain reaction that ultimately killed these cells. They could not only bias where a memory would land in the brain, they could also kill off the cells holding onto the memory at will. Josselyn and her team found that eliminating the small fraction of brain cells involved in forming a fear engram erased an animal's fear responses, providing evidence that the emotional memory was no longer present in the brain.

When their paper came out, Christine Denny was a graduate student in Rene Hen's lab at Columbia University. The members of the Hen lab also studied the neural mechanisms of learning and memory and used the Josselyn study as a launching pad to fine-tune our tools for shutting down a memory. Denny's PhD project was fascinating: by introducing genes that contained "instructions" for making green florescent molecules in mouse brain cells and installing optogenetic machinery, Denny discovered that she could see *and* inhibit a memory momentarily. Her work had altered the genome of a mouse so that the brain cells that held onto a memory would automatically glow green and could be shut off, rather than killed, at any time. Neuroscience's tool kit had become so sophisticated that it was no longer a matter of *can* we manipulate memories but a matter of *what kind* of memory and *for how long*.

The "on" switch for memory is what Xu and I discovered with our research: activating hippocampus cells housing a given memory was enough to activate the memory itself. The "off" switch is what the Josselyn lab had discovered: killing the brain cells housing a memory erased the memory itself. So Xu and I artificially activated a memory in the Tonegawa Lab, Josselyn and her team removed a memory from the brain, and the Hen team toggled memories off for moments at a time. All of our labs shared a vision of finding an "on/off" switch for memory, and the field of memory manipulation now officially had both.

———

In 2013, a few years into our friendship, Xu said he wanted to run some "big questions" by me. We were walking on a beach in San Diego, taking a much-needed break from a neuroscience conference, and procrastinating about going back to the apartment we were sharing with a dozen of our fellow lab members.

"Is there any memory *you'd* want to erase?" he asked me. Whenever Xu went out of his comfort zone to ask me something personal, I savored the chance to connect with him more deeply.

This took effort. I was tired. It turned out that tag-teaming multiple presentations about engrams during the day, while rooming with party-chasing guys at night, was a crash course in sleep deprivation and hangovers. I was hooked by his question because I could tell he was in a spiral of memory "what if's," and it was a chance to have Xu join me on a less-traveled, more winding memory lane of mine. I thought about his question and smacked my dried-out lips against the salty sea breeze.

Perhaps it was a defense mechanism of mine to tread gently into more serious conversations, but at this point in my life I had the knee-jerk tendency of making some quick wiseass quip in response to these kinds of questions. I couldn't help but toss in a flippant reply.

"The 2007 Super Bowl. The Patriots are still undefeated in my head—and that will forever be my truth." They had, of course, lost what otherwise was a perfect season.

Xu was so focused on the concept of memory erasure that he took my response at face value. Now that a couple of labs had conducted complementary studies that pointed to some incredible possibilities for the future, he was Promethean in his eagerness to shine the fire of memory manipulation for all of neuroscience, and what's more, he was ready to unpack every possible consequence of manipulating any kind of memory. "What's the point?" he asked. "Wouldn't erasing a memory mean you'd experience it all over again?"

His seriousness nudged me to be a bit more pensive. Xu and I found a quiet spot on the beach to eat the tacos we'd brought with us and continue thinking through the idea of editing out a moment of our past.

I asked Xu if he'd want to erase anything, and his response was simple and honest. "No, I don't think so. The good and bad are me," he said, pointing affirmatively to his chest. He didn't elaborate. He stared out at the waves, lost in his thoughts. I can only guess what kind of good and bad memories he was recalling.

I looked down at my feet buried in the sand. After a long silence, I told Xu there was one memory that I didn't know if I wanted to erase entirely but I did want to erase my feeling of helplessness tied to it because, to me, accepting this feeling in its entirety meant that I had failed at helping the world in any meaningful way like my parents hoped I would.

I told him that I had dreamed of running in the Boston Marathon since elementary school. My dad would say to me after his soccer games, "Running is the same in every language." He was a lifelong runner, and I became one too. Just as I had to practice an hour of piano every day, I also had to clock in at least thirty minutes on the treadmill, five days a week, in middle and high school. But unlike piano practice, which initially felt like a preposterous amount of theoretical physics homework for my fingers, running felt like an escape from everything. It helped that my dad somehow ran ten miles daily, six days a week, and then played soccer on Sundays, so I've always had someone to look up to who consistently did the impossible. Running brought me closer to him.

Throughout college and graduate school, my friends and I—and basically every student in Boston—partook in its city-wide tradition on Marathon Monday: we would celebrate by starting a party at eight in the morning and then take to the streets to watch in reverence as the runners sprinted past our cheering section. Boston's race had a respectful mirth around it.[2]

I hopped into the Boston Marathon on April 15, 2013. "You're not almost there!" read one of the signs held up near the beginning of the course. Every mile put graduate school on pause. *If I could run this many miles*, I told myself, *the following years of my PhD would be a walk in the park.*

The last twenty minutes of the race were as blissful and beautiful as they were exhausting and painful. My arm was draped over the shoulder of a college friend who had hopped in the race to run the last mile with me and help me limp my way across the finish line. At 2:35 pm, we finished the race and were headed toward an Italian restaurant just a few blocks away on Newbury Street.

"How do you feel?" she asked.

"Food," I exhaled.

At 2:49 pm, we turned left off Boylston Street and heard a blast. We froze. The thousands of people around us froze too, as we all tried to make sense of what was happening.

"What the hell was that?" someone behind us asked, in panic.

"I think a manhole just exploded," said another runner.

It felt like someone had taken control of a volume knob for the entire city. First the blast, then total silence. A few seconds later, a second blast turned everything into a deafening, skyward-bound echo.

During that echo, everyone transitioned from freezing to formulating an escape route.

"Just keep walking . . . just keep walking," I repeated to my friend.

People in the crowd started sprinting away from the finish line and, in that moment, my mind split in two. While my eyes darted all over, desperately seeking safety, I was also hyperaware of every sensation: the tension in my friend's shoulders when we put our arms around each other, the wobbling in my thighs as we speed-walked toward her apartment nearby, the ice-cold prickling in my fingertips. I remember repeatedly wondering what would have happened if I had finished the race just ten minutes later. What if I had talked a little more to the volunteers at each mile when I stopped for water? What if I had skipped a few training weeks and ran a slower pace? What if my ride to the marathon course was late that morning? What if, what if, what if? Sometimes the water of memory turns caustic and leaves scars, not grooves, in our neuronal riverbeds. This kind of engram blisters and quivers, restlessly, in the mind.

Now, on the beach in San Diego, Xu asked, "Would you ever want to run it again?"

My toes were cold and my palms were red and throbbing from clumping up the wet sand into dense balls of *what if*.

"You're frigging right I would," I snapped, still adrift in my own memory. "I replay so much of that race in my head. It feels like all of my training led to surviving that day rather than enjoying it."

Xu leaned back on his hands and spoke as if he were narrating his own life. He had a tendency to enunciate most words, and he emphasized them even more when he was making a point.

"Everything you live through is part of who you are. *Everything*. Sometimes you get caught in a loop, wondering what you could have done differently, but the reality is that you can never undo the past. You can only *re*do it."

He continued: "You finished the marathon, but it sounds like your marathon memory hasn't finished changing you. Maybe it won't finish until you run it again, or maybe it won't ever finish. I think both are okay."

Xu tapped me on the shoulder. "I'm glad you're here today."

"I'll let you know when I run it again," I said, thankful for his brotherly wisdom. Sometimes nothing from the outside world hurts as much as a memory from within, but a friendship from this outside world can help heal it.

Should we ever fully control memory? I thought about this while looking out at the ocean, imagining it'd be like peeling a rainbow one arc at a time to study each wavelength. Would we ever truly want to peel back the episodic bits, the emotional pieces, the semantic details of any given memory? The thought experiment was loaded with the weight of endless consequences. Xu and I imagined erasing the episodic and emotional components of my memory of Marathon Monday. Would I choose a full-blown neutralization of an experience, my episodic and emotional memory wiped like in *Eternal Sunshine*? Or would I just want to wipe out the emotional components, like in Josselyn's paper, and leave the experience itself intact and untethered from feelings? We could even make it so that what remained was only a semantic memory, one without dimension and without experience, just the leftover knowledge that I had run a race that ended in a bombing. This thought alone made me uneasy.

We pushed the thought experiment further. If I had my memory of *only* the 2013 Boston Marathon erased, then it follows that the memory of running the race—my euphoria of crossing the finish line and all the anguish that followed—would indeed be gone. But there's a very practical constraint to erasing any memory, which is that no one lives in a vacuum. I'd still live in a world that fully understood all the events and implications of that day. With information as ubiquitously accessible as it is now, how could I possibly avoid finding out what happened? This information would find its way back to me one way or another, and then I'd have to start a new process of coming to terms with the truth: I went through something so formidable yet chose to undo my conscious access to the experience. Such a nuanced approach to erasing specific memories would lead me to facts first before going through the associated feelings. It'd make all too palpable the difference between knowing something and experiencing it. The idea of reliving a terrorist attack, of going through an emotional episode that found its way back into my brain (and regenerated itself with symbiotic power) sounded unstoppable. It would be like a mental myth of Sisyphus, which I had no interest in and which finally meant I agreed with Xu. The point wasn't to undo but to redo, somehow.

When we got back to the room, I cracked open a beer and polished it off before Xu was done brushing his teeth. Then I started on a second one to decompress. While Xu enveloped himself with blankets and dozed off, I opened the window to listen to the Pacific waves, wondering—hoping—that recalling my memories of Marathon Monday was a good thing for my brain.

In the twelve years since 2013, I have often thought about the power that memories have over our lives and which ones are truly worth preserving. We all have catastrophic collective memories seared into our brains: a space shuttle exploding, planes flying into buildings, bombs going off in trains, school shootings, natural disasters, explosions at a finish line. And we all have our personal ghosts that come from a macabre past: family adversity, a traumatic birth, a frightening death of a beloved pet, a car crash, abuse of any form. Like Xu mentioned, these memories bend, twist, and ultimately form who we are. The ways in which they can be altered are not limited to optogenetics or some fancy new technique.

Science equipped us with memory suppressors that can warp how we recall an event, intervening with the brain while we're reexperiencing something. Many of these suppressors don't even require the complexity of altering the genetic makeup of brain cells or flashing lights into the brain; rather, they involve regimented changes to our actual behavior. Recall from Psych 101 how fear conditioning experiments work: a subject is given a neutral stimulus (such as an auditory tone) followed quickly by a noxious stimulus (such as a mild shock), so that the subject learns to associate the two. But if, over time, we were to give the neutral stimulus repeatedly and *not* deliver the noxious stimulus, this process would lead to "memory extinction," in which an organism learns to uncouple the emotional responses attached to a given neutral stimulus. Memory extinction is a process and applies to both negative and positive memories: it can occur through repeated exposures to the cue so that the organism now learns that the cue is no longer predictive of the negative or positive outcome to which it was once tied.

The concept of "extinction" is a core component of exposure therapy, and its underlying neural mechanisms can give us new leads for treating disorders of the brain, including PTSD and addiction, in which the unwanted return of debilitating emotional states is a core symptom. Exposure therapy changes our behavioral patterns and our sense of control over the stimuli and situations in the world that bring about fear, for instance, with the goal of reducing the incapacitating components of fear. This kind of therapy can help us overcome our fears and anxieties in a safe and controlled manner.

This process is called "extinction" not because the fear memory is *gone*, but because the subject suppresses its fear of the neutral stimulus itself. In other words, extinction is a type of learning by which an organism comes to know that a once-aversive environment or cue is no longer a threat. It's like when I made the brilliant decision early on in graduate school to go through a breakup with my girlfriend at Crema Café, my then-favorite breakfast place in Harvard Square. I used to go there daily to get a toasted peanut butter, banana, and honey sandwich, along with an iced coffee, and call my parents and wish them a good morning. But after the breakup, even just walking by Crema felt like an emotional roundhouse kick to my hippocampus. I had to force myself to walk back in, day after day, and begrudgingly continue my routine of sandwich, coffee, and morning phone call until I extinguished the I-hate-myself, Steve-get-over-it emotional responses.

The science of memory extinction adds a new layer to our understanding of how to disentangle a given emotion from its associated episode. In 2011, at the University of Puerto Rico, neuroscientist Greg

Quirk and two of his lab members, Demetrio Sierra-Mercado and Nancy Padilla-Coreano, conducted a series of experiments to study the similarities and differences between memory and its extinction. The team infused a drug into the rat brain to inactivate neural activity in two brain areas that they hypothesized were involved in fear memory recall and its extinction: the prelimbic cortex and the infralimbic cortex. (In humans, these are related regions located just behind our eyebrows, embedded in the prefrontal cortex, which is also involved in fear recall and fear extinction.) Quirk's lab found that quieting activity in the prelimbic cortex substantially impaired the recall of a recently formed fear memory, while silencing the infralimbic cortex impaired its extinction. This meant that the prelimbic cortex needs to be functioning healthily for a memory to be recalled; and the infralimbic cortex needs to be functioning healthily for extinction to successfully take place. The researchers had discovered both a way to prevent the return of fear and to impair the suppression of that memory, which have become major focuses of clinical studies in the last decade. Their study points to these areas, the prelimbic and infralimbic cortices, as potential therapeutic targets in humans with anxiety disorders or PTSD.

Within only a few years of the memory reactivation studies using optogenetics that I conducted with Xu, neuroscientists Ossama Khalaf, Johannes Gräff, and their team in Switzerland found that by using optogenetics they could reduce, or extinguish, the emotional responses of an old memory—which are harder to suppress than recent memories because the durability of memories can change and travel throughout the brain with the passage of time, as we saw in chapter 2. In other words, old or "remote" memories become more resistant to extinction and sometimes even strengthen their emotional hold over the brain as time goes by, morphing into something stubborn and stuck in its ways. All you have to do is think of the most embarrassing moment you've ever experienced (like the time you, let's just say, hydrated your pants in kindergarten, which was only noticed by the *entire* classroom when you walked back to your seat, which definitely didn't happen to me). Thinking about your kindergarten moment may bring back feelings of shame or regret, or both, despite being decades old.

Yet, neuroscience has successfully tackled the problem of suppressing even these old memories. By reactivating the cells holding onto the old memory, the Swiss team had demonstrated a kind of *internal* exposure therapy: they forced the memory to come back online, and it could now become disentangled from an associated emotion. Again, think of your most embarrassing moment, but now let's say someone could artificially reactivate this memory repeatedly until it no longer brought about its mortifying feelings—the embarrassment would detach itself from the moment and perhaps become a memory that we shrug off and view as character-building.

Neuroscience could now manipulate memories by targeting several brain areas, including the cortex, hippocampus, and amygdala, which gave researchers multiple innovative ways to change their impact on the brain. Our mental time machine, it turns out, has more than one button to jump-start it or to turn it off.

———

The recollection and extinction of memories are dynamic processes—they are concurrent interactions between the old and the new, between that which has been learned and that which is being learned. However, this wasn't always established in memory research. In the early 1960s, many scientists thought that, once formed, memories are fixed or "consolidated" in the brain, as I talked about in the last chapter. The accepted ruling was that memories are immutable, taking up permanent residence in the palatial confines of the brain. Then, in 1968, a group of scientists at Rutgers made a discovery in rodents that went against such dogma in memory research. The group showed that memories were not etched in stone but were flexible when recalled.

They first had rodents form a memory and then recall the memory the following day. The animals were then split into two groups: a control group and experimental group. In the control group of subjects, the brain was undisrupted when recalling the memory, which meant the memory could be *rest*ored safely in the brain again, a process referred to as *re*consolidation. Memories have to be tucked back in the brain after we wake them up from hibernation. But it's in this groggy, volatile phase that they're most vulnerable.

In an experimental group, the scientists electrically disrupted the brain's activity while the rodents were in the act of remembering, to test if they could interfere with the memory somehow. It's as if you pulled a book of memory out of its library, but someone tripped you while you were trying to read its contents. Everything gets scrambled in the shuffle. Against all dogma, the scientists found that disrupting the brain's activity while it was recalling a memory ended up dramatically impairing the memory—so much so that the animals now had amnesia.

It took nearly thirty-five years, but just as we were ushering in the new millennium, Karim Nader, Glenn Schafe, and Joe LeDoux at New York University were ushering in a fresh way to think about the neural mechanisms underpinning reconsolidation. The researchers knew that forming memories requires the production of new proteins and that these proteins help stabilize the neural changes that give memories their durability. The understanding at the time was that, once stored, the memory was concrete in the brain and no longer required subsequent neural modifications. But they discovered that the act of *recalling* a memory puts it in a susceptible state at the cellular level. It becomes physically vulnerable.

To get itself out of this vulnerable state, the memory must be stored again, which, like when it was stored the first time, requires the production of new proteins. To go back to our library analogy, it would be like checking a book out for a temporary amount of time, but once we're done reading, we have to put in some effort to actively return the book to its shelf so that it can be read again later. The scientists found that the *re*consolidation of a memory is in fact dependent on protein synthesis: administering protein synthesis inhibitors blocked the *re*consolidation of the memory and produced amnesia. Taken together, these findings indicate that remembering, like learning, is an active process. A memory at the forefront of the mind is a capricious thing.

While we can learn much about the cellular mechanisms of memories in mice, mice are not humans. This is a scientific fact. It's all the more amazing, then, when some aspect of cognition, like reconsolidation, shares similar biological properties across species. It took only a decade for this work on mice to be replicated in humans. In 2010, neuroscientists Daniela Schiller, Marie-H Monfils, and Elizabeth Phelps at New York University successfully updated fear memories in humans with non-fearful information by taking advantage of the reconsolidation window of memory. The scientists first had their human subjects form an emotional memory in the lab and then brought them in the next day to recall the experience.[3] If the subjects recalled the memory on the second day and then tried to extinguish it within six hours, this *permanently* erased the emotional responses associated with the experience. However, if the subjects recalled the memory and tried to extinguish it *after* six hours, the memory was left intact. This means that memories are modifiable but only if they are recalled and intervened with during that six-hour window.[4]

Why six hours? While the jury is still out, many believe that this is the time it takes for cells to build new proteins and reestablish the connections needed to reconsolidate a memory in the brain. If we interrupt the brain while it's hard at work trying to restore the memory—and this interruption can be in the form of extinction, direct activation of the brain, drugs, and more—then the entire process goes kaput and the memory breaks down. In the Schiller study, the fear memory was absent in the human subjects even a year later, while the other memories formed around the same time were still intact. As the team concluded: "These findings demonstrate the adaptive role of reconsolidation as a window of opportunity to rewrite emotional memories, and suggest a noninvasive technique that can be used safely and flexibly in humans to prevent the return of fear."

The researchers go on to suggest that reconsolidation allows old memories to update each time we retrieve them. In other words, a memory reflects our last retrieval of it; the memory can never be an exact account of the original event. This is the beauty and the pliability of memory itself. The brain, it seems, is a volcano of memories which form into rock when inactive and become molten when recalled, never to harden in the exact same shape twice. This updating has one ironic consequence: as Schiller once put it, the memories that are the most real, the most accurate, *are the ones we never recall.*

Memory updating offers us a way to "pluck from the memory a rooted sorrow". Leave it to Shakespeare to acknowledge this possibility of manipulating memories to alleviate psychological pain: "Canst thou not minister to a mind diseased / Pluck from the memory a rooted sorrow," laments Macbeth. "Raze out the written troubles of the brain / And with some sweet oblivious antidote / Cleanse the stuffed bosom of that perilous stuff / Which weighs upon the heart?"

Memory updating became personal for me rather quickly in graduate school. Shortly after the marathon bombings—and just months after I opened up to Xu about my feelings linked to the traumatic event—I began frequenting a brunch spot on the same sidewalk where the bombings happened so that I could attempt to update my experience of visiting Boylston Street. I wasn't trying to forget; rather I was trying to extinguish and rewrite what the memory meant to me. I knew that before my brain could adapt, it first needed to let these memories resurface. I recognize my sense of carelessness and desperation by trying to apply what I'd learned as a scientist, and all the studies I'd read in graduate school, to my own life. I mean, in practically every science fiction movie where the protagonists experiment on themselves, it ends up being a cautionary tale. "And that's why you don't become your own experimental subject, folks!" Not surprisingly, then, the process of trying to reconstruct my feelings associated with Marathon Monday were much more painful than I imagined.

At first, when I returned to Boylston Street, I could hear explosions, feel the ice-cold prickling on my fingertips, and see people running in every direction all at once. It was like fast-forwarding through an episode of my life where the only feeling associated with it was the helplessness I feared, all while a chaotic sensory experience from within left me temporarily stunned until the memory passed. These memories have that kind of paralyzing power.

My first few times back on Boylston Street also came with literal gut feelings that told me something terrible had happened here. I felt a burning sensation in my midsection, as if I had swallowed hot coals, while I tried to scribble down any details from memory of that day. My body simply did not want to be there, but I'd let these memories flow back in their entirety, hoping that one of those times would come with a bit of acceptance of what happened.

Xu said that I couldn't undo the past but that I could redo it, and I tried to redo Marathon Monday for about a year. There are, of course, scientific paths, like exposure therapy, to disentangle emotions from their episodes, but I took a more direct route with an all-too-human coping mechanism. Rather than undo or change an emotion, I numbed it.

The brunch spot I would frequent had a bar, and I learned early on that having a few drinks could quickly muffle the cacophony of the past so that I could focus on reliving one piece of an experience at a time. The first source of comfort came from a single sensory sliver of a memory: the powdery, lily-of-the-valley scent of my friend's perfume that occasionally drifted toward me while we hurried away from the blasts. Smell is our oldest sense, evolutionarily, and it taps directly into our hippocampal memory machinery. To me, the memory of this scent automatically signaled safety because it meant that I had a friend with me, despite all the uncertainty around us. It was a reminder that there was a *we* inherent to my memory of that day.

My brain was searching for a meaning that perhaps it wasn't ready to understand. I wanted to know *why we* survived. Sure, everyone has an expiration date. It's perhaps the single most important semantic fact that I know. But why was our date pushed back? My futile, ongoing search for an answer usually coincided with ordering more drinks, until the very process of remembering stopped for the day. If I continued to try to remember, I would break down crying. It's not that I didn't understand death. It's that I didn't accept it, especially having experienced an episode that let me know how suddenly it happened to people around me and how, without warning, it could have happened to me too. I was outraged that death was even a thing, let alone a thing I had no control over. *It is not fucking fair that we have to die.*

For over a year, I became my own experimental subject on Boylston Street, hoping to tilt the scales of neural activity away from trauma and toward relief. Feelings are data too. I can only speculate that repeatedly coming back to the finish line changed my emotional responses by altering the balance of activity between numerous circuits and reconsolidating the memory itself. These visits were some of the hardest things I'd ever done, and drinking got me through, so to speak, every single one of them. Recalling my memories of Marathon Monday and writing about them helped—extinction and reconsolidation partly worked—but it never got rid of the acute fear infused into my remembrances of April 15. That day produced a kind of anxiety that couldn't be remembered away. As Xu said, perhaps undoing such a memory wasn't even the point. It's part of my experience and identity, whether I like it or not. My memories are me, regardless of what kinds of memories they were in the first place.

So I persisted. Over time, walking down Boylston helped me reframe April 15 into something I at least had more control over. Each time I thought of crossing the finish line, a second, more harmonious memory became inextricably linked with it. That second memory was of April 17 when Rene Rancourt, a legend in the Boston area, sang the national anthem before a Boston Bruins game, just as he had for over thirty years. Halfway into, "What so proudly we hailed," his voice faded out and the sound of 17,000 voices filled the hockey arena with the kind of reassuring warmth my memory of Marathon Monday needed. It was a reminder of hope. It was my friend's perfume all over again. I was learning to surround April 15 with the kind of memories that gave me the slightest glimpse of the answer to *why* we survived. Marathon Monday was as much a somber reminder of life's fragility as it was a celebration of life persevering. Both thoughts were true to me; I am not sure I want to erase the memory that taught me so much. As the novelist Robert Louis Stevenson wrote, "It is the history of our kindnesses that alone make this world tolerable." I believe the same applies to making memories tolerable too.

————

While extinction and reconsolidation help suppress the emotional components of memories, how can we ever know if a memory is really gone? We know from studies that it's possible to target a memory's underlying emotional and/or episodic part, and I know from my own work with mice that activating the cells harboring memories can jump-start recollection. But we can also look to popular culture for some guidance on this journey. Christopher Priest's novel *The Prestige* (the basis for the 2006 movie) provides some food for thought:

> Every great magic trick consists of three parts or acts. The first part is called "The Pledge." The magician shows you something ordinary: a deck of cards, a bird or a man. He shows you this object. Perhaps he asks you to inspect it to see if it is indeed real, unaltered, normal. But of course . . . it probably isn't. The second act is called "The Turn." The magician takes the ordinary something and makes it do something extraordinary. Now you're looking for the secret . . . but you won't find it, because of course you're not really looking. You don't really want to know. You want to be fooled. But you wouldn't clap yet. Because making something disappear isn't enough; you have to bring it back. That's why every magic trick has a third act, the hardest part, the part we call "The Prestige."

If you'll entertain me for a second, please allow me to reinterpret this speech within the context of memory:

> Every great memory manipulator consists of three experimental acts. The first is called "The Memory Trace." The scientist shows you something ordinary: your memory of a birthday party, the family dog, or your father's voice. They show you an impression, a physical trace of your experience, in the brain. Perhaps they ask you to test if a memory is indeed real, objective, truth. But of course . . . it may not be. The second act is called "Memory Erasure." The scientist takes your ordinary memory and makes it do something extraordinary. It's somehow, seemingly deleted, and now you want to look for the original version . . . but you won't fully find it, because you can't really look. You forgot how to remember. But you are being fooled by your memory. You don't give up yet. Because making a memory disappear isn't enough; you have to bring it back. That's why every memory manipulator has a third act, the hardest part, the part we call "The Amnestic Trace."

What if memories simply can't be erased? What if their supposed absence is one of the greatest magic tricks the brain can play on us? In fact, emerging research points in this direction, revealing that memories are never really truly *gone*; they may just need to be "hotwired" back to life. The brain—the grandest magician of them all.

———

Around the time our first paper on reactivating memories came out in 2012, our engram team began to grow. When Tomás Ryan walked through the doors of the Tonegawa Lab, I noticed three things: he was approximately one thousand feet taller than me; he had really slick sunglasses on; and he was wearing a black coat that made him look like Neo from *The Matrix*. Kind of badass, if you ask me. He was standing in the coffee room, clearly waiting for someone.

"Who's the new guy?" I asked Xu.

"He's interviewing to work in the lab as a postdoc fellow, to study how the brain remembers and forgets," Xu said.

When he interviewed with Susumu, Tomás said he wanted to study how cells that are active during the encoding of a memory are modified when a subject attempted to retrieve that memory. This was particularly important to him because the unsuccessful retrieval of a memory is, by definition, amnesia, and if we could now bridge how the brain *encodes* memories with how it *retrieves* memories, then perhaps we could pinpoint what happens when retrieval fails. His experiments centered on testing if amnesia meant that memories were no longer present in the brain, or if it meant that memories were still there but inaccessible. In the library of memories, does amnesia destroy the books themselves and get rid of any trace of a memory, or does it incapacitate the librarian who now can no longer retrieve each book? Tomás outlined his experiments with Xu and me, as they relied on verifying our first study on artificially reactivating memories. Luckily for us, he was earnestly collegial at heart, and Susumu hired him shortly thereafter. Team X officially had another member.

When Tomás started in the lab, there was an unpopular theory that had been around since the 1960s: perhaps in some circumstances, amnesia means the librarian really *did* temporarily go on a break, and the books were still intact. This theory stated that the production of proteins lays down the neural material that allows the *retrieval* of memories rather than the *storage* of memories. The theory was unpopular because of the thousands of studies reporting that blocking proteins would prevent memories from being encoded, but these studies never actually tested if that memory was truly gone from the brain. Tomás's brilliant hunch was that memories are more permanent in the brain than we expected, and our inability to recall certain memories is a result of a retrieval failure. "Memories are the most unbreakable thing that I know of," he said.

"This guy *really* knows his shit," Xu told me. He was once again right: Tomás was a walking encyclopedia of memory research. Xu and I would learn that he first got into memory research after a nasty accident gave him amnesia about the entire incident. These memories, according to him, have been completely lost since then. What remains is just the semantic knowledge of the event, without any episodic or emotional details.

In our lab, Tomás now had everything he needed to test if he could bring a memory back from amnesia. To do so, he started by challenging a fundamental assumption in memory research: that learning induces changes in neurons, and these changes require the production of proteins, which are the brick and mortar of the brain, to stabilize the memory over time.

When an animal experiences an event—a new environment, a new food, a new littermate—blocking the production or *synthesis* of new proteins seemingly blocks a memory of it from forming. For instance, we can place an animal in one of our conditioning chambers and let it experience an aversive event. Immediately following the event, we give the animal a drug to block the production of proteins in the brain. Then, on the following day, we can place the animal back in the environment, and it'll display evidence of amnesia: it'll act as though it hadn't experienced the negative event. Blocking protein synthesis immediately following the formation of a memory blocks the memory itself.

Tomás's ingenious experiment went like this: he first installed ChR2 in the mouse brain cells that were active during the formation of a fear memory and then immediately administered a drug to prevent the production of new proteins in the brain. When the animals were placed back in the environment where they formed the fear memory, they behaved as though nothing negative had happened, as expected. They were amnestic. However, when Tomás optogenetically reactivated the brain cells processing a fear memory, the animals immediately froze. Despite exhibiting amnesia, he could nonetheless artificially bring a memory back. Tomás had awakened a dormant memory that we thought was forever lost.

How were the animals recalling a memory, even though the very memory itself had never been allowed to stabilize? How could a hiker who leaves no footprints ever be found? We know that the memory, the hiker, had to leave *some* kind of mark since all experience leaves a physical trace, but just what this trace looks like was the million-dollar question. It turns out that when it comes to the brain, the hiker may not leave behind footprints, but the hiker does leave behind a trail of crumbs so they can retrace their steps and make the return trip.

The first act in the magic trick of memory manipulation is to see and activate a memory as Xu and I had done with Project X. The second act is to erase a memory, as Sheena Josselyn had done in Toronto. And this was the third and final act of the magic trick: a true *Prestige* in the brain. Tomás and his team's data demonstrated that as long as we have access to the cells that processed the memory in the first place, then we have a target for retrieving it. His paper's title reveals the magic in scientific terms: engram cells retain memory under retrograde amnesia.

Tomás's findings soon sparked headlines around the world: "Neuroscientists reactivate 'lost' memories in mice," "Amnesia patients could recover their lost memory," "Bringing memories back from the dead: science, not fiction." He invited Team X over to his house one night to toast the milestone that we had just witnessed. I learned more about good scotch and Irish politics than I could handle and woke up the next morning on his couch with a bit of my own amnesia.

Neuroscientists could now play hide-and-seek with memories, but what about memories that we've lost due to time and the natural act of forgetting? If you think back to the first memory you have, how old were you? Many studies report that prior to the age of three, our cognitive machinery hasn't sufficiently matured to be able to hold onto the daily barrage of experiences we go through during that phase of life. Our inability to retrieve memories prior to age three is termed *infantile amnesia*, and we think it is the only type of amnesia we all share.

A few years after Tomás's study, neuroscientist Paul Frankland, his graduate student Axel Guskjolen, and their lab at the University of Toronto published a groundbreaking paper that answered whether or not memories formed in infancy were lost forever or if they could be brought back. Frankland and his team found the brain cells in the hippocampus that were involved in forming a fear memory *early* on in an animal's life, tricked the cells to have ChR2 installed in them using the same genetic tools Xu and I used, and then asked if they could bring a memory back later on in the animal's life. The team optogenetically stimulated the cells that encoded the memory in infancy but did so when the animals were now adults. To everyone's amazement, even though the animals displayed infantile amnesia (they could not naturally remember the infantile memory), the researchers could both optogenetically reawaken the memory when the animals were adults and even retrieve the memory at virtually any age. "Recovery of 'lost' infant memories in mice," proclaimed their paper's title. "The finding that the engrams still exist in the brain in a latent or 'silent' state might explain how these forgotten memories continue to influence our thoughts and behaviors as adults," Axel told me over coffee one day.

The presence of amnesia may not be evidence of an erased memory; rather, it is evidence of a lost memory waiting to be found. A general truth about the scientific process, popularized by Carl Sagan, is also true in the context of memories: "Absence of evidence isn't evidence of absence."[5]

A reoccurring theme in modern neuroscience is that memories indeed are more present and unbreakable than we believed. There are, of course, many ways to seemingly break a memory: people living with drug addiction are more susceptible to having amnesia because of the effects of drug abuse on memory; the sleep-deprived brain powerfully impairs memory or flat out renders it naturally irretrievable;

Alzheimer's disease, perhaps the most devastating memory disorder we know of, produces a profound amnesia for the episodes of life that one holds so dear. But we're now learning that even though memories may not be naturally accessible, they are artificially accessible. Since Tomás's study, scientists have consistently been able to artificially retrieve memories thought to be lost. As long as the brain is intact, it appears that memory remains artificially retrievable, even if a subject is presenting all the symptoms of amnesia.

We're all born with amnesia, and many of us will leave the world with amnesia; that much is true. But together, Tomás Ryan's findings and the findings of the Frankland lab form a majestic case of the amnestic memory trace. They give us hope that engrams we thought had vanished are in fact living in a mental shadow, waiting to be illuminated. (The plot of the show *Severance* relies almost entirely on this neuroscientific framework for how memory works!) Maybe the memories that Josselyn erased actually went into quiescence and remained hidden somewhere in the brain, waiting to emerge again at some point. Maybe my memories of Marathon Monday still contain hidden details that I haven't recalled since April 15, 2013, waiting to spark unexpectedly, for a spotless brain is impossible. This possibility alone keeps me going back to Boylston Street.

And yet, even if memories are impervious and feel like accurate accounts of our past, how would we ever know if they are *true?* Enter the phantasmagoria of the real and the false when it comes to the act of recollection, the subject I'll discuss next.

This I know: everything I have learned and have shared in this chapter, and in this first part of the book, on the mysteries and revelations of memory, culminates in a life lesson that I carry with me—the episodes we experience in the world, layered with the emotions we feel, and imbued with the facts that we accrue into knowledge, come together to form a keystone of our core as human beings. They all have a reason for being. And for me, memory's multifaceted magnificence takes on true meaning with a single thought that I hold close to my heart: Xu will always be my friend.

4

More to Remembering
Than Truth

A WEIRD thing about the brain is what it chooses to remember. Fifteen brief minutes in 2013 changed my life, yet I remember a grand total of absolutely none of it. If you'd like, you can experience those fifteen minutes too—the video of Xu and me giving our first TEDx talk together was a huge opportunity to share our science with the world. I know from watching the video that it happened, and it looks like I may even have been enjoying myself. But whenever someone asks me what it's like to give a TED talk, the most honest answer I can give is, "I don't know."

Why can I not remember something we practiced every single day, twice a day, for sixty days straight? I do remember that when Xu would mess up in practice, he'd tense his arms up in front of him while getting back on track, like a conductor trying to readjust the orchestra's pace. I saw Xu the kid who didn't like being the center of attention and Xu the scientist who strived for perfection. I witnessed a conflict playing out in front of me—Xu's conflict of who he was and who he wanted the world to see.

I also remember being borderline possessed by practice: when I was in the shower, I'd rehearse our talk. When I was out for a jog, I'd rehearse our talk. When I was trying to wind down for bed, I'd rehearse our talk. When I was out on a date, I'd rehearse our talk. When I was rehearsing our talk, I'd rehearse our talk. I was going nuts—this was our chance to tell the world about the dawning era of memory manipulation, and we both wanted to do our best.

I remember checking in at the Seaport World Trade Center in Boston with Xu and getting our name badges, which read, to our amusement, "Brain Inceptors." Just outside the auditorium we found a dimly lit practice room and went over our talk. Approximately ten rehearsals and an hour of hyperventilating later, we got the word that we were next. My brain encoded the next moments in crisp detail. As I took my place on stage, it felt like someone had injected vinegar into my legs. I looked down past my shaky knees at the red TED circle. In my peripheral vision, I saw where Xu was standing to my right. The beaming auditorium lights created enough lens flares to blur out individual faces in the audience. *Thank god I don't have to make eye contact with anyone*, I thought. I heard the murmur of the audience still to silence.

The next thing I remember, the auditorium was filled with applause and Xu and I were saying thank you together before marching off stage. As we approached our seats, we looked at each other and shrugged.

"It's a little different than lab meeting," Xu laughed.

"My face feels like someone put a heat lamp up to it," I said.

Piecing together what it was like to be on stage with Xu during our talk has been an exceptional exercise in getting nowhere. To recall the specifics, I talked to Xu, to audience members, and to other speakers present that day. I've watched this talk repeatedly and searched for the tiniest detail that might trigger my memory of the experience and re-animate my first-person presence on stage. Sometimes, brief glimpses of experience seem to come to mind: as I'm looking out to the crowd, the audience is laughing at a joke I made about a former relationship; Xu is yelling that our experiments worked; I can feel prickly beads of sweat on my back. But are these flickers of the real memory tucked away in my brain, or am I accidently inventing some of my own account by repeatedly revisiting those fifteen minutes so many times?

Every time I close my eyes and focus on walking up to the red circle with Xu, I get so excruciatingly close to having the ensuing fifteen minutes come into focus. But it's more like I'm taking a candle to Plato's wax tablet (which we discussed in chapter 1): the closer I get to that engram, the more its shadowy surface melts into something new. As much as I don't want to accept it, I doubt I *actually* remember any part of being on stage at all.

The theme of this chapter is simple, but its implications are anything but: whatever is remembered gets changed. Because of this, our quest to manipulate memories will go from pretty neat to kind of weird. Just as memories shape-shift with age, we'll see that they also shape-shift each time they are remembered, warping into an entirely new shape with new experiential contours. How can we manipulate memory if it is already constantly changing? Can we use the malleability of memory as a tool to help us alter memories in the lab?

Neuroscience in the twenty-first century reveals that memories are dynamic reconstructions of the past and not just static stamps of experience. From DNA to cellular activity and all the way up to the contents we consciously bring to mind, memories are rebuilt each time they are called upon. As I keep trying to recall my memory of being on stage with Xu, I'm inadvertently etching into it details that never were. So while memories are pastiches of experiences built from events that happened, they are all too easily filled with fabrications that turn the past into a psychological fiction. For better and for worse, false memories can be the most convincing lie our brains tell us.

In the lab, Xu and I were spellbound by this malleable nature of memory and initiated Project Inception to understand how false memories are created in the first place. Within a year of artificially activating a memory, we went on to successfully implant a false memory in the brain, and the neuroscience community quickly began to manipulate—to activate, erase, implant, and restore—all sorts of memories across the entire brain. What was once the thing that only nature could do to memory was now the thing that scientists could do at will.

———

Hollywood has evocative ideas about how realistic memories can feel. In the film *Blade Runner 2049*, the film's protagonist, K, visits Dr. Ana Stelline—a memory fabricator, who can make and view memories in other people. On a personal mission to find out if one of his memories is real, K asks Dr. Stelline to look into his mind as he recalls the memory. "Think about the memory you want me to see. . . . Let it play," she says to K, as she is seated, looking into one side of a V-shaped, snow-white microscope, while he stands in front of the other side. She views his memories, presumably by deciphering the activity in K's brain and stitching together a composite glimpse of his past experiences. She tears up.

"I know it's real," he says. "I know it's real."

Dr. Stelline doesn't respond. K tightens his face and screams desperately, kicking his chair over, realizing that his beloved childhood memories are actually false.

While Hollywood has free reign over depicting memories on screen, viewing stable, permanent memories in the lab gets a bit trickier. In a landmark paper, neuroscientists Leon Reijmers and Mark Mayford at Scripps Research Institute asked their rodents to do the same thing that Dr. Stelline asked of K: *think about the memory you want us to see.* The researchers developed genetically modified mice in which amygdala cells that were active while they formed or recalled a fear memory became permanently "tagged" by a fluorescent protein. If an amygdala cell glowed red, then it meant it was involved in the formation of the memory. And if a cell was involved in retrieving a memory, then it would glow green. But if a cell was active for *both* the formation (red) and retrieval (green) of a memory, it would now glow *yellow*. With these tools, the authors were able to capture snapshots of the brain as an experience entered and left its mark. Amazingly, yellow cells were in abundance.[1] This study was instrumental to engram research because the researchers discovered *a stable neural correlate of a memory.* Perhaps what Dr. Stelline viewed through her microscope was what the researchers viewed in their animals: the crystallization of the seemingly ephemeral material of memory.

Within a few years, neuroscientists Kiriana Cowansage and Mark Mayford discovered that memories can be artificially modulated in *numerous* brain areas, not just in the amygdala as the Josselyn lab had done or in the hippocampus as Xu and I had done. They could even reactivate a memory without a hippocampus, and the artificially activated memories looked very much like real memories in the brain, since both recruited similar neural circuitry. It's why K's memories felt so real. As we'll see below, our own brains have a hard time distinguishing between what is false and what is real when it comes to remembering the past.

Indeed, we all have false memories. Consider September 11, 2001. Many of us born before the mid-1990s could describe what we were doing, where we were, and how we felt on hearing the news. These are the kinds of memories that really stick. We'd easily and confidently describe, for instance, watching news outlets replay footage of the first and second plane hitting the World Trade Center. The problem, however, is that the footage of the first plane hitting the North Tower wasn't readily available until the next day; we installed that new detail into our memories of the day before. The amalgamation of shared experiences can form a collective memory for an entire population, but even these collective memories can be embedded with misplaced details.

But some of our memories *we know are true*. Let's consider three harmless examples of real memories that we're certain are accurate. After, we'll go back to some examples of false memories and highlight the differences.

Finish the final line of the celebratory Queen song:

We are the champions!
We are the champions!
No time for losers 'cause we are the champions
Of the _____

Or, fill in the blank for the following legendary Star Wars quote:
"_____, I am your father."
Finally, in Disney's *Snow White*, what are the first two words that the Evil Queen says while looking into the mirror?
"_____, _____ on the wall."
If you guessed, "Of the <u>world</u>," "<u>Luke</u>," and "<u>Mirror, mirror</u>," then, guess what?

Wrong. Wrong. ALL WRONG. Queen's "We Are the Champions" ends with "We are the champions!" and that's that. Darth Vader never says, "Luke, I am your father." He says, "*No*, I am your father." The Evil Queen never says, "Mirror, mirror, on the wall," but says, "*Magic* mirror on the wall." See how easy that was to get you to recall a false memory? Voilà! Inception.

It's called the Mandela Effect. The name comes from the popular belief that Nelson Mandela died in prison and that many people remember watching his televised funeral shortly thereafter. Mandela, in fact, died nearly two decades after being freed from prison. Why are our memories so naturally susceptible to modification? Since the Mandela Effect owns enough real estate in our brain to warrant a mortgage, let's take a closer look at what modern neuroscience has to say about the biological basis of false memories.

Memories are biological attempts to store information for future use and to use this information to make predictions about what we may experience next. The brain is constantly trying to stay one step ahead of the present. Because it's in nature's nature to be imperfect, many memory-like mechanisms are subject to errors, but these errors also have their advantages. For instance, DNA is inherently embedded with errors after each round of replication, but these errors help to diversify the genetic pool over time. Your skin and bones have types of memory, too: scars and fractures are physical traces that the slings and arrows of outrageous pains in the ass happen, but skin and bone can grow back, albeit noticeably different than the original copy. Where there are errors, there are opportunities for our biology to adjust and strengthen itself.

Updating old information with the new can be advantageous, and our technology has already figured this out: "Save As," a feature in Microsoft Word, updates the original document into as many iterations as we want while preserving some essence of the original file. Our culinary prowess is no different: Chicago-style pizza, dare I say, *still contains bits of pizza information*, but with crusty new dimensions compared to a floppy New York slice. And we can spend hours on Instagram updating photos by using filters to change the original image into a new masterpiece. Likewise, our memories can be edited post-production.

Our brain's gift of being able to hold onto a remarkable amount of information for flexible future use comes at the expense of never being a perfect record of the past. Even at the molecular level, the formation of a memory leads to small cuts in DNA that are quickly repaired in order to stabilize a memory. From the genome to behavior, memory is mutable, and that's both a bad thing and a good thing. False memories are linked to the wrongful sentencing of hundreds of people—to lifetime imprisonment, or to death—based on faulty eyewitness testimony. Distorted memories can surface during a therapy session and lead us to misremember important details of a past experience. Memories, real or not, powerfully influence how we understand the present day, be it the news we encounter, the information we receive, or the lens through which we view the world.

But the malleability of memories also opens up the possibility that they can be changed in a therapeutic manner. Heraclitus was on to something when he said, "No person steps in the same river twice." Likewise, no person experiences the same memory twice. Despite their lasting presence, memories are rivers that ebb, flow, and reconstruct the neuronal riverbeds that structurally support various streams of information.

Elizabeth Loftus, distinguished professor of psychology at the University of California, Irvine, and Daniel Schacter, human cognitive psychologist at Harvard University, have studied human memory and its underlying neural basis since the 1970s. Their pioneering work on memory is second to none.

Xu and I interviewed Loftus in 2014, and she told us that, when recalling a memory, "[people] are taking bits and pieces of experience from different times and different places and combining them together to construct what feels to them like a memory." In one of her most famous studies, Loftus ran an experiment where she had more than 1,200 participants study slides depicting an automobile accident (see the 1978 paper, "Semantic Integration of Verbal Information into a Visual Memory"). After going through the images, each participant was given either information that accurately described the car accident or misinformation related to the event. Misleading the subjects to incorporate false details into their memory was all too easy: many could be tricked to misremember details surrounding the car accident, such as what kind of traffic signs were present, and to misremember the speed of the collision. In the 1978 paper, Loftus calls it the misinformation effect and concludes, "Overall, the results suggest that information to which a witness is exposed after an event, whether that information is consistent or misleading, is integrated into the witness's memory of the event."

Her experiments underscore just how easily outside information can sneak its way into our memories. A follow-up study in 1995 showed that it was possible to convince 25 percent of participants that they had been lost in the mall at some point as children, when they hadn't, and these participants went on to fill in vivid details of an event that never took place (see "The Formation of False Memories"). As a result, Loftus has advocated for the findings of neuroscience and psychology to inform and reform our legal system, especially because the particulars of a re-called experience can be easily altered. It's not to say that we can't always trust memories; it's to say that we must be careful when considering memories as purely objective testimonies of the past.

Building on the idea that memories are fallible, in 2001 and 2004, Schacter's lab published two monumental studies seeking to answer the following questions: What is the neural basis of false memories, and can the brain distinguish true from false memories? Schacter's lab used what's perhaps the best tool we have to eavesdrop on human brains working in real-time: functional magnetic resonance imaging (fMRI), which researchers use routinely to noninvasively peer into the human brain while it performs its daily cognitive miracles.[2]

Using this state-of-the-art imaging technique in 2001, Schacter's lab recorded the signals a brain gives off when it's recalling events that did and didn't happen. To begin, his team had to make people form real memories. And since we're all friends here, let's have some fun: I'll give you a real memory for free, and you don't even have to Venmo me! Read over the following list of words just once, and then I'll give you a test that I bet you can't pass. But you have to promise not to cheat. Ready-set-GO:

Water
Fish
Boat
Swim
Stream

Lake
Beach
Sun
Canoe
Paddle
Splash
Sea

Alright and now here we are, ready for the test to end all tests? Of the following five words, which ones did you *not* see:

Water
Engram
Boat
Ocean
Patriots

If you're like me, then you chose Engram and Patriots. If you're also like me, then you may have missed that the word *Ocean* didn't in fact appear on the list. This list had a theme, i.e., bodies of water, which biased you to believe that the conceptually related word *Ocean* appeared in it. It's a mild type of false memory but one that's easy to induce in the lab setting by presenting related words, stories, sounds, images, or movies. Called the Deese–Roediger–McDermott paradigm, it's a classic task in cognitive psychology often used to generate false memories in humans.

When Schacter's team had their subjects recall true memories (e.g., *Boat, Water*), the fMRI registered a large increase of activity in the hippocampus. This made sense: we've long-known that the hippocampus is important for forming and recalling personally experienced events, and these results reaffirm its role recalling segments of our past. However, when subjects recalled a false memory (e.g., *Ocean*), the fMRI *also* registered a large increase of activity in the hippocampus.

It turns out that the same area that helps us recall our worlds of the past also helps us misremember these worlds. Speculations abound as to why. Perhaps it's because the act of remembering involves conjuring

up a mental representation of a previous experience that itself is a con-
taminated copy of the past, since our biology is spectacularly imperfect.
Or perhaps it's because the biology underlying false memories wasn't
much of a burden (it didn't consume many resources) for natural selection's merciless razor to trim away. Maybe false memories are the
price we pay for having rich recollections, and the latter proved more
advantageous than the former. It might indeed be the case that false
memories put us at an advantage and serve a double-edged purpose by
misrepresenting the past. By doing so, they enable one of the most magical mental feats humans are capable of achieving: imagination. But before pulling the rabbit out of the cranium on why we have false memories, we have to visit one more study to set the stage.

After Schacter's 2001 study, the team now turned to the question
posed previously: given that false memories exist in the brain, can we
identify a type of neural fingerprint that's unique to what we misremember? To answer this question, the group showed multiple shapes, some
similar and some different, to a human subject in an fMRI machine.
During the recall phase, when some subjects misremembered seeing a
shape, many brain areas came online that were *not* online during the
accurate recall of a true memory. In other words, recalling a false memory produced its own pattern of neural activity, which was different
from a real memory. The difference between a true memory and a false
memory was located in an area of the brain that's pivotal for processing
visual information, namely, the visual cortex.

If you want to have an existential crisis tonight, ask yourself this:
Given that you are your brain, and given that your brain is capable of
distinguishing true from false memories, why do false memories still
feel so real to you? Shouldn't you be able to discriminate between true
and false memories given that your brain can? Yes, the visual cortex can
discriminate between true and false memories but, as Schacter concludes in his team's paper: "the sensory signature that distinguishes true
from false recognition may not be accessible to conscious awareness."

Schacter describes the aftermath of these studies in his splendid book
The Seven Sins of Memory. When someone publishes a paper that claims
to have found a neural fingerprint capable of differentiating true from

false memories, people tend to listen. They also email and call a lot, especially with regard to eyewitness testimonies. Can we simply put someone in an fMRI machine and get a neuroscience version of a lie detector test? Can we look at their visual cortex and have science tell us whether they're lying or not? The answer, as of now, is no. The findings of many fMRI studies are based not on individual brains but on *averages* across dozens of brains. Currently, an fMRI study of a single brain can't predict whether someone is recalling a false memory or a real memory.

As Loftus's and Schacter's work suggests, your brain is full of real and false memories. It indeed can tell the difference between the two. But you cannot. Said another way, you can tell the difference between true and false memories, *but you don't know it*. Telling the difference requires other people. As we've learned, the information needed to distinguish true from false memories relies on an external observer and on unbiased observation with tools like an fMRI. It relies on someone who isn't confounded by being in your shoes. That is where our field is going: one day in the future, you'll be able to tell the difference between true and false memories by letting go of control and relying on someone (ideally a neuroscientist) other than yourself.

The work of Loftus, Schacter, and their colleagues reminds neuroscientists and psychologists that there are puzzling twists in how false memories are made. Everyone's memories are protean. Even people with a "super memory," who remember every detail of their lives, are just as prone to false memories as us mere mortals. And people living with certain kinds of dementia are sometimes even more susceptible to false memories, for reasons that remain unclear.

Which brings us back to our original question: Why do false memories exist in the first place? Psychologists have speculated that false memories might be a price we pay for having imagination. It might be the side effect of being the creative, limitlessly meditative species that we are. Several cognitive psychology studies have shown that people confuse memory with imagination all too easily. As Loftus and Schacter note, "Simply imagining an event that might have occurred in one's personal past can increase confidence or belief that the event actually occurred." When the researchers used fMRI to test this, they observed that the brain displayed similar patterns of activity for both recalled and imagined events, especially in the hippocampus. Extraordinarily, much of the same neural machinery that helps us remember and reconstruct the past also helps us imagine the future. Past, present, and future converge on many brain areas, and one of them is the time machine that is the hippocampus. The inextricable link between memory and imagination will bring us fully up to speed with modern neuroscience research and theories that we'll more fully flesh out in chapter 6.

————

So far we've learned that our mighty memories are metamorphic. With Project X published and out in the world, Xu and I began brainstorming a sequel project that focused on *warping* memories. While celebrating the publication of our memory reactivation paper, we barhopped our way around Times Square and deliberated all the conceivable ways we could change an engram from within the brain. The billboards turned into kaleidoscopes as I exhaled what must have been tequila vapors, visibly condensing in front of me, into a carefree mist. I could feel my breath steam onto my beard and told Xu he shouldn't shave until we could get another paper out.

"Absolutely not," he chuckled, "I'd look like a leopard."

I checked my phone and it was almost three in the morning—obviously the perfect time to plan our next project, despite us looking like two newborn antelopes trying to walk for the first time.

I could give you every hypothesis and every scientific rationale as to why tinkering with memories is important, but it would only be part of the truth. Xu and I found ourselves most excited about the brain when we were unrestrained by dogma, when we felt like we could truly pluck questions from the tree of science fiction and ground them in experimental reality. The name of this new project, drunkenly bestowed by Xu and me, was all too fitting. We sat down at an Irish pub, ordered wings, and Project Inception was underway.

Xu always had a pen on him. He grabbed a napkin, and we began discussing a set of plans that we had brainstormed with Team X back at

MIT. We drew three squares in a line: Box A, Box B, and Box A again. In these experiments, we'd first place a mouse in Box A. While the mouse was forming a memory of this environment, we would genetically trick the brain cells active during the formation of this memory to respond to brief pulses of light because we'd installed ChR2 in them. This way, we'd now have control over what we'll call a "neutral" memory, since nothing good or bad happens in Box A.

Next, this same animal is placed in Box B. If we shoot light into the brain and activate the hippocampus cells that are still processing their memories of Box A, then the animal should "think" that it's in Box A. And here's the million-dollar experiment: we optogenetically reactivate the memory of Box A while the animal is in Box B, and at the exact same time we simultaneously give the animal a mild foot shock. It's an attempt to link the artificially activated engram of Box A with the foot shocks occurring in Box B. If we can get the animal to "think" that it's in Box A while receiving foot shocks in Box B, then when we place it back in Box A, it should now show evidence of fear, even though nothing bad ever happened here. If we successfully implant a false memory, then the mouse should freeze in an environment that is actually safe.

A few weeks after our New York trip, we were heading into the holiday season and ready to test our ideas for Project Inception. By this point, I had found a groove in graduate school. Thanks to Xu's training, I could perform brain surgeries like a pro and run optogenetic experiments routinely. For once, I had a sense of control in science. The lab was quiet the day before Christmas, and I was alone in our testing room performing the hour-long experiment that we had planned in New York. My parents were outside waiting for me, parked adjacent to our building's main entrance and texting me that everyone else was waiting back home to start Christmas lunch. My brother and sister—both lawyers who speak legalese and not memories—were also texting me to hurry my ass up and get my mouse shenanigans done. Apparently food waits for no one in the Ramirez household.

The six animals I was working with had already undergone all but one step of the protocol Xu and I outlined: we made the memory for Box A activatable by light; we reactivated this memory in Box B while the mice were given the mild foot shocks; and now we could go through the final step. This was it. Would they actually freeze when I put them back in Box A?

As the first mouse entered Box A, its ears perked up and its head quickly darted left and right. Within a few seconds, the mouse scurried to a corner of the box and remained motionless for the duration of the test. One after the other, each mouse froze in Box A. I couldn't believe it. I'm pretty sure I dislocated my right shoulder quietly fist pumping as each mouse successfully recalled something they had never experienced. The experiment—and this basically never happens in grad school, I reminded myself—actually worked. Before I could register what I had witnessed, the most important hour for Project Inception was over. My hands were shaking and I scribbled notes to send to Team X. I could not wait to tell everyone.

I shot off a quick email with a graph of the data to Team X: "MERRY FREAKING CHRISTMAS!"

Xu immediately responded with a congratulatory message. I'd choose this memory, and all the feelings in it, over trying to remember my fifteen minutes on stage any day.

———

After Projects X and Inception, Xu and I felt like we were a part of a renaissance in memory research. Manipulating memories became rather ordinary, and we were caught right in the middle of this fertile subfield of neuroscience. Fittingly, it took just a little over a year for our false memory research to beget a fresh direction in the Tonegawa Lab. Team X welcomed several newcomers who all became comfortable mastering the art of memory manipulation. Our engram group had turned into a force of nature that was fueled by the fusion of friendship.

With many duos now emerging in the lab, two gifted neuroscientists, Roger Redondo and Josh Kim, teamed up right away. In addition to activating memories, they wanted to do something entirely new. They wanted to scramble the contents of memories and completely switch their emotional components by taking advantage of the pliable nature of recollection. Specifically, they wanted to turn a positive memory into a negative memory, and vice versa.

Like the rest of Team X, Roger and Josh spearheaded experiments that demanded punishing shifts for months at a time. They didn't shy away from putting in grueling hours to finish their project at a breakneck speed. (Which is not to say that this is a glorious and healthy way of doing science—*it absolutely is not*.) Their project required two steps:

Step 1: Activate a positive memory. Roger and Josh knew they could reactivate negative memories based on our previous work, but now they needed to reactivate a positive memory. What does a mouse find inherently rewarding?[3] (If your answer is *cheese*, close this book and go take a long glance at the shame that is the mirror.) They turned to one of nature's most potent rewards: social interactions. They began by placing mice together, finding the brain cells in the hippocampus that were involved in forming this positive memory, and tricking those brain cells to respond to light.

Step 2: Turn the positive memory into a negative memory. Intuitively, this step can happen to humans instantaneously. Imagine eating your favorite tub of ice cream but then getting food poisoning—you'll probably avoid cookies and cream for some time. Or imagine kayaking in the picturesque Caribbean seas, and out of nowhere a wave tips you over and you find yourself drowning (because the aquatic treadmill that is the ocean requires you to have the kind of upper-body stamina that I 100 percent did not have that day). Our brain handles such near-instantaneous experiential volatility by layering the newest coat of emotion onto a given moment.

Roger and Josh's hypothesis was audacious: emotional memories from within the brain can be rewritten with experienced emotional events from without. To test this, they *artificially* reactivated the positive memory within the rodent's brain while the rodent was *naturally* forming a negative memory. Amazingly, the brain cells that once activated a positive memory now had the *new* capacity of activating a negative memory. And the reverse held true too: they could force cells that once activated a negative memory to now become cells that were capable of activating a positive memory. They had successfully gained a "bidirectional switch" for positive and negative engrams.

Despite working together on this research, many of us inadvertently sacrificed an immeasurable amount of mental and physical health. Out of shame, I didn't tell Xu or the rest of Team X how much I stumbled in the months following our Project Inception paper. It was becoming increasingly hard to separate my work from my sense of purpose. Each phone call, interview, lecture, argument, and late-night musing almost always focused on my research. I was becoming my work, which meant that my science was personal. At the same time, I was starting to feel like I finally didn't suck at doing research. While this didn't mean that I was giving up having a life outside of lab, I let the line between my work and my perceived self-worth blur, despite Xu's warnings. I started to feel like the only way I could matter in science was to continue discovering new things at an even faster pace. More and more, I was turning to partying as a pressure relief valve. Turning happy hours into late nights let me escape briefly from the stress of the lab and the expectations I had placed on myself. I was careening through graduate school without a seatbelt. For now, I recklessly stayed the course and willfully turned a blind eye to any signs telling me to slow down.

In its purest form, good science anywhere is good for science everywhere. A year after Roger and Josh's discovery, neuroscientists from Kaoru Inokuchi's lab at the University of Toyama, Japan, further advanced engram research by artificially linking two distinct experiences in the brain. In a paper published in 2015, the experimenters were able to optogenetically activate the memory of an environment and the sensations of pain to link the two and thereby form a qualitatively new memory. Around the same time, neuroscientists Felicity Gore and Richard Axel at Columbia University found that artificially activating cells that produced innately negative or positive responses could be used to impart that meaning to something as bland as a tone. The researchers could now give emotional meaning to otherwise neutral stimuli—a sound that initially didn't matter to the mice now evoked cellular and behavioral responses indicative of emotional memories. And in a mind-blowingly provocative finding from a paper aptly titled, "Memory formation in the absence of experience," Gisella Vetere, Paul Frankland, and their lab in Toronto were able to write a memory from *within* the brain without the animal having to ever experience the event.

"Memory is coded by patterns of neural activity in distinct circuits. Therefore, it should be possible to reverse engineer a memory by artificially creating these patterns of activity in the absence of a sensory experience," they postulated. The memory they implanted consisted of artificially created links in the brain. They did this by stimulating brain cells that produced physiological responses that were innately rewarding, while simultaneously stimulating brain cells that evoked a neutral sense of smell for mice, such as the smell of an orange. By activating both types of cells at once, the scientists made it so that their rodents now showed a strong preference for orange-scented stimuli, even though they had never experienced these before! And they could do the reverse: if they activated innately *fear*-producing brain cells while simultaneously stimulating cells that evoked the smell of orange, then the animals would now show a strong aversion to the scent, despite never having experienced the smell before. Their artificially created memories even looked similar to real memories in the brain, so by all measures they had inscribed a de novo experience that the animal perceived as a memory, even though it was of something that never happened in the external world.

Xu and I learned that, at any given moment, there were suddenly multiple Project Xs and Project Inceptions going on in neuroscience, each capable of artificially creating false memories, linking separate experiences, and eliciting emotional responses instantaneously. The highlight reel of these projects in contemporary engram research has been as technically impressive as it's been biologically profound. For instance, researchers at the University of Washington developed a "holographic optogenetics" approach to stimulate individual brain cells one-by-one that harbored positive or negative memories, which meant it was now possible to link the activity of single neurons to their corresponding memory as well as the behavioral change the memory causes, all with cellular pinpoint accuracy. Researchers in Paris, France, deconstructed a memory into its moment-by-moment components, such as when a mouse is experiencing a stimulus or when it's producing a freezing response, and they identified the cells responsible for each moment. These cells, it turns out, harbor distinct aspects of a memory, and yet stimulating them is enough to trigger full-blown memory recall! As if that isn't exciting enough, two labs based in Toronto and New York artificially activated memories in mice to get the animals to actively make decisions based on the environment the mice "thought" they were in, such as digging for a food reward or avoiding a specific part of a context, even though the mice were in completely neutral environments that had never had these experiences associated in them. Altogether, within just the first two decades of the twenty-first century, neuroscientists began to extract memories and emotions from the brain, alter their contents, and then measure how these changes affect the cognition and behavior of a subject.

Shortly after Roger and Josh's paper on "switching" the emotions associated with a memory was published, Xu and I went to the Muddy Charles Pub, a staple hangout spot at MIT, to celebrate and plan our next and final project. Reactivating memories was so 2012, creating false memories was so 2013, and scrambling up the emotions tied to a memory was so 2014. We were looking forward to having another year of working together and now asking if memory manipulation could be used in the context of psychiatric disorders. Xu and I had reactivated a negative memory, we had incepted (or implanted) a false memory, and now we wanted to break into the brain again and activate a positive memory to *fix* a brain from within itself. After a pitcher of beer and a few hours of outlining what the experiments would look like, Project Zero Mouse Thirty was underway.

We knew our experiments hadn't yet attempted to use memory manipulation as a therapeutic strategy. We also knew we were short on time for this project. Xu was about to begin the onerous process of applying for faculty positions, and I had started to plan my great escape from the Tonegawa Lab to launch my independent career as a neuroscientist. I was hoping to graduate around the time that Xu would leave the lab, so this was the last project we'd launch at MIT. Despite the high-profile publications, the TEDx talks, the interviews, the glitz and glamour of it all, I wish I had known that this was the last project Xu and I would ever work on together.

5

The Antidote from Within

EVER SINCE I can remember, I've been captivated by superheroes—the comic books, the movies, the mythology of each universe. As a kid, every superhero contained some power I wish I had: the ability to fly, to read minds, to teleport anywhere in the world. They were fictitious ideals. Over time, my interpretation of superheroes became more grounded as their humanity became more apparent to me. I began to identify more and more with the hardships of their journeys: the desire to fit in, a personal struggle with the past, a calling to help people. The more I paid attention to the hero's journey, the more I saw that it reflected the wishes and struggles of all people. Superheroes are commentaries on the human condition because they embody an ideal balanced by some very human vulnerability.[1]

The closest people I have to superheroes in the real world are my parents. When my dad snuck across the border to the United States, my mom stayed back in El Salvador to take care of my older brother and sister. She was their everyday shelter from the ongoing civil war. The country's extreme socioeconomic imbalances led to a decade-long war between the government and guerrilla groups, which would go on to take over 75,000 lives, displace half a million people, and force an additional half a million refugees to flee the country. My mom was among those who fled. This was how she spent her twenties, long before I was born: during the morning, she walked my siblings to school, then she went to a neighboring village to get supplies to bring back to La Reina, her village, where her parents asked her to start a makeshift corner store to help with the family income. During the night, my mom stayed huddled close to my brother and sister—they all slept in the same room—and she comforted them amid the rapid-fire midnight clamor that signaled impending bloodshed.

Once a month, my mom walked to the local telecommunications center, *Antel*, where she waited until an operator, who managed the two phones available on-site, signaled that a call for her had come through. My dad phoned in at the same time each month to update her on their plan to bring the rest of the family to the States. After two years of monthly phone calls, one conversation brought the news she had been waiting for: my dad said he had made enough money to bring her, my brother, and my sister across the border. My mom had achieved the heroic ideal of protecting her children long enough to give them a second chance at life. The anxieties of living in a war would now transform from daily experiences to memories of a life that once was.

Superheroes put in the kind of work that the rest of the world won't. In Boston, my mom worked as an animal technician at Harvard for fifteen years, performing the often-tedious duties of cleaning and changing hundreds of mouse cages daily for biology researchers. Through years of working with other bilingual Latinos, she overcame the language barrier and her default shyness—just enough to comfortably communicate with colleagues at work. The uncertainty of restarting life in a foreign land slowly turned into a comfortable morning routine: wake up, drive with my dad to Dunkin' Donuts, get dropped off at work by six in the morning, and at three in the afternoon go home to enjoy an early dinner with the family. Life had become predictable which, compared to life back in La Reina, meant wonderfully safe. Her memories of El Salvador transformed yet again into moments of gratefulness at the dinner table.

"*Tenemos que decir gracias a dios por cuidarnos todos los días,*" she would say reverently before a meal. "We have to say thank you to God for watching over us every day."

Every superhero's origin story is somehow connected to the struggle they will endure and overcome at some point in their journey. While my mom shared many memories of her childhood with me, the first time I learned that she was in yet another battle in her journey was on a cloudy June afternoon, midway through my time at MIT. The day could not have been more routine: she was in the passenger seat of our SUV, talking with my dad about our next trip to El Salvador. While stuck in a typical Boston rush hour, my mom became lightheaded, and her heart began beating abnormally fast, seemingly out of nowhere. Her hands were clammy, curling and uncurling as if grasping the air for some semblance of relief. The next few minutes, she said later, felt like a vice grip was constricting her throat.

My dad held her hand as they began breathing in and out together.

A few agonizing minutes elapsed in slow motion, as her breaths became willfully deeper. A bit of control was coming back. Her cheeks were apple red, and her hands had the kind of reflexive jitters that one gets after walking in subzero temperatures without any gloves. My mom had just had a panic attack—a bout of overwhelming anxiety that can instantly debilitate a person and entrap their mind with sustained tension and fear. She braved the full expanse of life, but the reality instantly sank in: even superheroes get hurt.

Feelings like anxiety are hard to understand because they're often invisible to everyone else. But they're part of everyday life. We all regularly experience bouts of stress and anxiety. What if the job interviewer doesn't like me? Should I cram tonight for my exam tomorrow? What should I say on my first date? Our biology often forces us to be prepared for multiple outcomes in the midst of uncertainty. It's healthy to care about these multiple outcomes because it encourages us to put in the work, to adequately prep for a given stressful event. And yet, sometimes the scales of stress become tilted to such an extreme that pathologies of the brain begin to emerge.

What I want to emphasize in this chapter is *not* what is sometimes believed in neuroscience: broken brains give rise to broken thoughts. No—the thesis of this chapter is that *a broken brain does not make a broken person.* To the contrary: I believe that a broken brain begins a search for healing. We are all broken (or healing) in our own way, and I view that as a unifying human bond we all share. Sometimes differences in mental health will appear foreign to others because we don't fully understand them. The tremendous variation in how any individual arrives at a state of anxiety, for example, highlights that our brain contains many winding roads that can ultimately converge on the same feeling. We all have our triggers in life, but what those triggers are depends on experience—on memory. When these differences impair our mood, thinking, behavior, and overall daily functioning, then they get lumped into a category; what's more, if observed impairments share similar features, then this category itself falls within a broader classification—that of a *mental disorder*.

To me, neuroscience shouldn't focus solely on "fixing" the brain but rather on simultaneously healing a person holistically and building a society in which our differences are supported and understood to improve our overall well-being. As Xu would tell me, while it is a brain that gets fixed, it is a person who is healed. When our well-being is in harm's way because a particular difference becomes a symptom that is persistently debilitating to our daily life, that is when we search for a remedy. As we'll see in this chapter, mental health is scientifically accessible down to the very cells that make the varieties of mental disorders possible; when these disorders are considered within the context of our biology and within the environment in which they exist, then we can truly achieve a full, empathetic understanding of what defines a human.

I believe we can reinterpret our inimitable internal characteristics as a mental diamond, unique in its shape, which we can rough-hew as we will. My mom's memories of her time in El Salvador form her mental diamond—one which I can peer into to see both glimmers of her core and bits of myself. Her memories, and the anxieties that come with them, have given her the kind of superhuman strength that helped her survive and live a fulfilling life. They have also produced some of the worst days of her life. There is therefore a patterned pressure that shapes our mental health: what is adaptive in its origins, such as a brief bout of anxiety to presumably avoid some danger, can collapse into a maladaptive condition with no end in sight or obvious treatment available. That is, unless we figure out how to nourish our mental health with the antidotes that modern neuroscience is beginning to provide.

It is when a hero is confronted with their whole self, when their wishes and struggles align to yield purpose, that they are free to triumph.

As we'll see, a combination of genetic predispositions and environmental risk factors underlie psychiatric disorders, such as anxiety, depression, and posttraumatic stress. Like a storm we can't control, our neural circuits sometimes unpredictably spark and thunder with enough force to leave permanent marks on the landscape of our brains. We can understand these marks by studying and embracing the neurological differences between people, which lets us come together as an empathetic society as we transform our concepts of mental health into a celebration of human diversity. I am uniquely me, you hold the world record for being you, and our humanity binds us.

Mental disorders do not define a person or usurp our humanity; rather, they underscore and nuance the vast heterogeneity existent across every single brain and offer a very private insight into our individuality. They are also not a choice. People can *think* away a psychiatric disorder as much as an athlete with a broken leg can *think* away the fracture. Both are physical, both require a physical understanding, and both require a physical approach.

So far, our quest to manipulate a memory has taken us from turning memories on and off, to creating false memories, to restoring those memories thought to be lost. Now I explore how these triumphs of neuroscience can be used to promote an organism's health. Memory manipulation is an exciting part of our tool kit for healing the mind, and the techniques for this kind of clinical application range from noninvasive strategies like cognitive behavioral therapy and clinically prescribed drugs, to more invasive strategies like direct stimulation of the brain itself with electricity and lasers.

Today, neuroscientists are redefining mental health disorders from a brain-centric perspective. Despite sometimes having sinister, Hollywood-inspired connotations, memory manipulation can be used to enable biological well-being. Xu and I embraced this prospect head-on in one last project together in which we artificially activated positive memories as a sort of antidote that, surprisingly, was enough to treat symptoms associated with anxiety and depression in mice. What would new treatments look like for the brain if memory was viewed not just as a facet of cognition but as a tool to restore health to an individual?

———————

As I was entering my last year of graduate school, I couldn't stop thinking about my mom's panic attack during that morning commute. Anxiety injects chaos into the mind, and no matter how much we tell it— sometimes *plead* for it—to stop, it persists and debilitates. I was set on applying our memory manipulation tools to tackle psychiatric disorders in general and to test whether or not artificially manipulated memories could restore order to the mind. My last project in graduate school would attempt to artificially activate positive memories to suppress the symptoms associated with anxiety and depression. It would be my most personal scientific endeavor, a very direct way for me to join the fight at my mom's side and to thank her for being my superhero. If my research could somehow inspire new therapeutic strategies that might be useful for alleviating these kinds of debilitating conditions, then my work with Xu will have gained an even deeper, more personally meaningful purpose.

Throughout history, psychiatric disorders were viewed as pathologies of the soul, mind, or body; they weren't always seen as disorders of the brain. The treatments for these pathologies were similarly wide and varied, including performing exorcisms in the Mesopotamian era to get rid of the demons causing depression, removing blood in the Greco-Roman era to balance our bodily fluids, spinning a person around in the nineteenth century to induce dizziness and increase circulation, and performing lobotomies in the twentieth century to "cure" a person.

Their cause and definitions were just as mysterious. Sigmund Freud proposed that imbalances between our conscious and subconscious were the root of psychological disorders. The German psychiatrist Emil Kraepelin, who many consider to be the father of psychiatry, developed a highly influential classification scheme to compile a list of common *behavioral patterns* of symptoms over time. He was one of the first to consider clusters of symptoms—lack of sleep, hallucinations, rapid changes in mood, for instance—as diagnostic criteria for a psychiatric disorder. Indeed, today's bible for diagnosing mental disorders, the *Diagnostic and Statistical Manual of Mental Disorders*, or DSM, owes its approach to Kraepelin's way of grouping symptoms together. The symptoms associated with a psychiatric disorder now had a collective manifestation with discrete names.

Between 1949 and 1958, serendipity kicked in, with treatments for mental disorders aimed directly at changing the balance of chemistry in the brain. Drugs that were meant to be used for sedation before surgery were recognized for their calming, antipsychotic capacity. Other drugs that were meant to treat tuberculosis were inadvertently found to improve patient's moods. Large doses of insulin and shock therapy were used to treat psychosis. It's as if everything that was meant for the body worked unintentionally for the mind as well. Within a decade, the brain was no longer a black box but a key player in producing pathological cognition and behavior. It was a true revolution—when psychology at last met biology, and together they became neuroscience.

As a testament to how real this revolution was, nearly every antidepressant, antipsychotic, and anxiolytic drug available today is an iteration of a drug that was discovered between 1949 and the early 1960s. By analogy, our technological devices today are also tweaked versions of their equivalents from the last century. This is because our tweaking *via science* produces tangible results. We're not carrying boulders for cell phones and computers the size of elephants anymore. They're in our pockets. And the doctor doesn't look at a broken toe and strap a patient into a contraption with pulleys and weights, only to then recommend a full leg amputation. We get an X-ray, maybe some surgical screws and painkillers, and then back to the comfort of our couch we go. Since humanity has refined the technological advancement of our smartphones and medical practices for our right toe, then why not apply similar societal and financial gusto to modern-day research?

We all stand on the scientific triumphs of the last century. It's now time we call for a neuroscience revolution for treating the brain and healing a person. There is hope because science works. Drug-based treatments can be effective, and researchers have painstakingly outlined their biological mechanisms of action; so it is no surprise that most of the currently available drugs reproduce many of the beneficial effects first discovered in their 1950s counterparts, as Steve Hyman—former director of the National Institute of Mental Health (NIMH)—points out. Such drug-based treatments are a means to restore health; and if we want bona fide cures, then the recipe is rather straightforward: fund researchers to do the science with a sustainable and forward-thinking infrastructure. That's how we got to the moon, and that's how we'll cure diseases. This way, we can launch another revolution and continue to use neuroscience as both a way of understanding our inner world and ridding it of all unnecessary suffering. Neuroscience is an international dialogue that can transcend all that divides us by tackling the problems that happen to unite us through shared biological adversity. In other words, disorders of the brain and body do not discriminate and neither should we. Every single one of them has a scientific solution that we can discover if we create a sociopolitical environment that permits the science to launch us into the next era of human exploration. With this exploration comes a deeper understanding of our biology, and that understanding enables the very flourishing that is the antithesis of disease. It is how we win against all that harms us.

Neuroscience indicates that the fluctuations we experience in our mental health are a result of measurable changes in the brain and behavior. The DSM classifies and sets forth criteria on defining psychiatric disorders based on changes in behavior. These changes come in the form of patterns and clusters of reoccurring symptoms.[2] This classification scheme for diagnosing psychiatric disorders is both useful and not quite enough. "To date, the diagnosis of mental disorders has been based on clinical observation, specifically: the identification of symptoms that tend to cluster together, the timing of the symptoms' appearance, and their tendency to resolve, recur or become chronic," Hyman notes. Yet, many contemporary researchers would argue that such a classification scheme is insufficient if it doesn't include *both* the behavior and the underlying biology.

Let's take one example. By DSM standards, two people living with generalized anxiety disorder each might display three of the disorder's six symptoms but share only one. By analogy, two patients might both have a fever, but one additionally has a headache and nasal congestion while the other is vomiting and bleeding from the eyes. The former has the common cold; the latter has Ebola. We treat these patients differently because we know that the common cold and Ebola present themselves similarly in some behaviors but with completely different causes. But when it comes to mental health disorders, we tend to lump patients together by diagnosing them with the same disorder if their behavioral symptoms are deemed similar enough. Classifying mental disorders based solely on irregular changes in behavior treats each behavior as a unitary entity that varies enormously in its biological origins.

In 2010, NIMH launched an experimental attempt to bring psychiatric disorder diagnoses into the twenty-first century: rather than starting with a top-down approach by using clusters of psychologically observable *behavioral* symptoms, they proposed starting with a bottom-up approach by using neuroscience to look at the brain itself. In contrast to the DSM, the Research Domain Criteria (RDoC) project classifies disorders on the basis of what's happening *in the brain* as opposed to solely relying on behavior. It demolishes what I call the myth of the mental monolith. The mental monolith is the idea that a psychiatric disorder can be described simply by one defining behavioral feature and that this feature maps onto one area or chemical in the brain. But depression is more than just a shortage of serotonin; dopamine is more than just the "pleasure hormone"; and the hippocampus is more than just memory. Each disorder is a result of staggering biological complexity, and the RDoC framework attempts to take such complexity into account.

As Tom Insel, also a former director of NIMH, summarizes in a 2013 article "Transforming Diagnosis": "While DSM has been described as a 'Bible' for the field, it is, at best, a dictionary, creating a set of labels and defining each. The strength of each of the editions of DSM has been 'reliability'—each edition has ensured that clinicians use the same terms in the same ways. The weakness is its lack of validity. Unlike our definitions of ischemic heart disease, lymphoma, or AIDS, the DSM diagnoses are based on a consensus about clusters of clinical symptoms, not any objective laboratory measure."

One of the goals of the RDoC project is to incorporate objective biological measures into how we classify and treat psychiatric disorders. In addition to studying behavior, the RDoC program breaks the brain down into "domains," into a set of neural systems and circuits, each involved in processing positive, negative, cognitive, social, and arousal-related information. These domains all have both human and animal-model analogs, so that studying an animal brain can still theoretically yield some insight into the human brain, and vice versa.[3] Importantly, they are starting points, and not definitive criteria, for evaluating an individual brain. When coupled with the DSM's descriptive, behavioral criteria, the RDoC criteria will be continuously refined as our tools and strategies for understanding the brain in the context of its environment evolve simultaneously.

The RDoC program has the advantage of viewing disorders of the brain as existing on a spectrum: each individual consists within a range of variability, from cells to behavior. An RDoC approach to anxiety breaks the condition down into the "negative valence systems" of the brain. These are the systems the brain uses to confront the endless number of scenarios that can cause us harm in the world. In this heightened anticipatory state, we may worry excessively, even when there isn't an obvious threat; nonetheless, our body responds by preparing for danger. The ways in which our body goes into its all-hands-on-deck mode cut across all levels of biology: stress molecules, like cortisol, flood the nervous system and engage cells throughout the brain that produce aversive states, such as fear, which produce changes in behavior that are simultaneously shaped by the present environment. Variation in any level and in any direction may produce a markedly different activity pattern and, therefore, brain state. Like a symphony orchestra tuning their instruments before a performance, the RDoC framework breaks down each section ("system") of the brain and listens for what may be out of tune first before making any necessary adjustments.

When we start to treat symptoms as products of neural activity, then the expectation is that precise physical patterns of neural activity will relate to specific thoughts or behaviors, and that these microscopic patterns in the brain will have unique physical features, which interventions can one day directly target. In rodents, for instance, optogenetic studies have been successfully able to worsen or alleviate very specific symptoms of anxiety, including changes in heart rate, rewarding sensations, and avoidance of threatening situations. The combination of the RDoC conceptual blueprint with targeted interventions of the brain would allow symptom-specific tailored treatments for any of the core components of a psychiatric disorder.

Helen Mayberg, of Emory University, and colleagues reported that it was even possible to *predict* which treatments would most benefit a person living with a psychiatric disorder. What they found in their experiments, led by Callie L. McGrath in the Mayberg Lab, was truly spectacular: by scanning the brains of patients with major depressive disorder prior to treatment, the researchers found that the patterns of communication between a handful of brain areas accurately predicted whether cognitive behavioral therapy (CBT) or medication worked better for treating patients—one pattern of activity pointed to CBT, while a separate pattern of activity pointed to medication.[4] As Mayberg points out: "All depressions are not equal and like different types of cancer, different types of depression will require specific treatments. Using these scans, we may be able to match a patient to the treatment that is most likely to help them, while avoiding treatments unlikely to provide benefit." Like Mayberg's study, the RDoC framework stratifies the tremendous heterogeneity of psychiatric disorders into discrete units in the brain, and then relates how these units contribute to a distinct behavior.[5]

———

Xu and I took a similar brain-centric approach for our newest project.[6] Could memory itself be artificially controlled in rodents, by tapping directly into the brain to restore neuronal and behavioral balance in a therapeutic manner?

Luckily, our project had a scientific precedent in humans. One morning, Xu pointed out an influential paper by psychologist Barbara Fredrickson and colleagues called, "The undoing effect of positive emotions"—namely, the capacity of positive emotions to undo the physiological effects that negative emotions have on the brain and body. The undoing hypothesis proposes that positive emotions can be used for more than just feeling good. They can be used to help us: get out of bed in the morning; pursue happiness; change how we think about and interact with ourselves and others; and counteract, or at least regulate, negative emotions. When human subjects were stressed and then watched movie clips that elicited contentment and amusement, their bodies rebounded in beneficial ways: their stress-induced increases in cardiovascular activity, for instance, returned to baseline faster than when they watched neutral or sad movie clips. Excitingly, this reveals a very real physical connection between feelings of positivity and their direct effects on our biology.

Xu and I wanted to further this work by testing for a potential therapeutic capacity of positive memories by jump-starting their biology from *within* the brain. We placed our animals in a box that had two small valves on separate ends: one that delivered sugar water when the animals licked it and another that delivered regular water. This is known as the sucrose preference test. Rodents normally prefer sugar water over regular water, the same way humans will typically find sugary liquids preferable to a bland liquid. On the other hand, rodents with depression- and anxiety-related behaviors tend to show a 50:50 preference. They show no preference at all.

As expected, the animals displaying anxiety- and depression-related behavior licked at each of the valves randomly over the course of fifteen minutes. As with Project X, all we had to do was hit a button that would turn our lasers on and optogenetically awaken a memory from within.
Click.
The deep-blue laser flickered throughout the mouse's hippocampus, waking up—*activating*—cells that held onto a positive memory. I remember thinking that our optogenetic stimulation was a fancy, high-tech Proustian madeleine, one capable of triggering the rich remembrance of things past.[7] If you'll entertain my romanticization of the moment: the mouse perked up immediately, as if a shudder ran from its brain to its body, and it began scanning the environment to decide

which valve to visit first. An extraordinary thing was happening. I imagine that the mouse felt the memory invade all its senses, strangely detached and with no suggestion of an origin, since the essence of these sensations was *in* the mouse as much as it *was* the mouse. And once the positive memory fully revealed itself within seconds, the now-motivated mouse inspected each valve with some sniffing, followed by a taste test. When it found the valve with the sugar water, the mouse started licking vigorously, so much so that it consumed as much sugar water as our control animals. In under an hour, Xu and I saw that reactivating positive memories restored our mice's behavior to a healthy baseline. Just as exciting, reactivating positive memories also turned on many areas of the brain involved in rewarding experiences and motivation. The key to reversing abnormal behavior was embedded within their positive memories all along. For as long as the laser was shining its sapphire radiance in their brains, the mice were motivated to keep consuming their sugar water reward. *All this from stimulating cells in the hippocampus.*

Said with less novelistic flare: the mice got a sugary treat.

In the following weeks, one of my gifted undergraduates Briana Chen collected a large empirical dataset for the project, and it came with an exciting plot twist: when she artificially reactivated positive memories twice a day, or "chronically," for about a week, not only did this permanently ameliorate symptoms we believed were associated with depression and anxiety, but it also *promoted* the growth of new cells in the brain. She had found a new way to force the brain to make more brain cells and replenish itself, while simultaneously and permanently restoring the animals' behaviors. Positive memories had both short- and long-term benefits, all the way from cells to behavior.

Inspired by the neuro-centric RDoC approach to treating the brain, our hope was that the biological potency of positive memories—like medications—could inform cognitive-behavioral approaches to treating disorders of the brain. This project was meaningful to me on a personal level: I thought of my mom and the idea that she might never have to experience the kind of crippling anxiety that robs someone of peace. Around the time our paper came out in 2015, neuroscientists Megan Speer and Mauricio Delgado at Rutgers University discovered that in people actively recalling positive memories decreased blood stress hormones and activated brain areas known to be online during rewarding experiences; in addition, as in our mice, the group later found in 2021 that recalling positive memories rapidly reduced negative emotions in a manner that lasted at least for months.[8] More recently, research led by Asya Rolls found in mice that activation of the brain's reward systems even improves recovery from a heart attack. The idea that we can use a "mind over matter" strategy can work if we acknowledge that the mind *is* matter and, therefore, the activity of the mind can be therapeutically targeted to modulate the very well-being of the body that contains it. Taken together, these findings underscore that savoring our positive memories—recalling them intentionally and really sitting with them as they replay in our mind—broadens our cognitive ability to problem solve, to cope with all sorts of stressors, and to manage our bodily physiology in a manner that promotes health.

Positive memories are some of the most powerful biological tools available in our brains. At home, my mom and I shared a treasure trove of them—one that we both remember is from the time when I was a teenager, and we were visiting her parents in El Salvador. One morning, my cousins, parents, and grandparents all walked down a hill behind the house my mom grew up in to go swimming in the village pond. My cousins kept egging me on to jump from a cliff into the pond, and my mom kept telling me I didn't have to. Like her, I was the opposite of an adrenaline seeker because, oh I don't know, maybe my innate biology was onto something, as "please do not free-fall to Earth" kept repeating in my mind. She could see that I was scared, and after a few minutes she suggested, much to my surprise, that we jump together. We held hands and tiptoed to the edge—*uno, dos, tres*—we were in the air! Moments later, we emerged from the water laughing in delightful disbelief at our newfound courage.

Neuroscience tells us that this memory has all the ingredients of life's dessert that make us feel good. From an RDoC perspective, my cognitive and valence systems are all interacting to produce the riches from this experience: the cognitive system enables the memory of jumping off of a cliff, which at first generated feelings of fear via the negative valence systems, which are now almost immediately counteracted by feelings of reward via the positive valence systems. What was once a moment of fear is now a memory of triumph with my mom. It's the only time I can remember when we both took a literal leap of faith, so we cherish the memory as an example of what our brains can achieve together. A million little life moments like these, packaged neatly into a million memories that we hold onto constitute the good stuff in life.

———

We've seen that positive memories influence our brain and behavior in useful ways. Forming them brings us joy, and remembering them brings us stability. Their beneficial use extends beyond individuals too: sharing positive memories with others nurtures a powerful connection, a bond, between people, especially if we know that one day we'll have to say goodbye.

On January 21, 2015, Team X was ready to celebrate Xu's last day at MIT. Xu had accepted a faculty position at Northwestern University, and his scientific vision was about to materialize: The Liu Lab.

We decided to go to one of our usual haunts, less than a mile from campus: Green Street, a restaurant with an unassuming brick façade, which gives way to a pleasantly fancy and dimly lit, two-floor, white-walled interior. We had a reservation for more than two dozen lab members.

"Burgers and beer is exactly how I thought I would spend my last T-Lab outing," Xu remarked, while chuckling in his staccato way.

He and I left the lab early to cash in on some quality time before the crowd arrived, and Xu commented every block on some piece of architecture or small park he was going to miss about Cambridge. He walked like a daydreamer, with no obvious destination. He'd see the negative space in the world and wonder out loud about its purpose—how the intersecting columns of a half-built building would inadvertently frame two people working behind it, or how the leaves on a tree gave different shapes to the clouds beyond them. I was going to miss his calm demeanor and his perceptiveness of the world around him.

We sat next to each other in the middle of a long, brown, rectangular dinner table as Team X trickled in, along with the rest of the Tonegawa Lab. Xu and I took a selfie and gave a toast: to being grateful for having the first-world problem of studying engrams. Immediately after our toast and while fixing his black, thick-rimmed glasses, Xu downplayed how much his hard work had paid off and said, underneath, he was still just a student doing science. He embraced that humble, hardworking approach to life.

Throughout dinner, Xu showed Team X his favorite pictures of his time in the T-lab (positive engrams galore!): one photo was of him on his back drinking a yellow Gatorade after hiking up Mt. Monadnock in New Hampshire with a few lab members; another photo was when we went apple picking in Western Massachusetts, and we'd had a contest over who could throw a rotten apple the furthest; and one showed the lab's thirteen-person dragon boat team which, through some miracle, actually won a race against Harvard. Every few minutes, I noticed that Xu covered his teeth when laughing, despite having had his braces removed recently. He hid his new smile out of habit, and I found it endearing. I was going to miss those small insecurities.

Team X and the entire Tonegawa Lab united for an evening of celebrating the fact that Xu had reached his escape velocity from the lab and was now living on the edge of what is and what will be. As the night progressed, I became more and more eager to give him my going away present because of what it symbolized. That morning, I had gone to buy him a gift and was completely stumped. I wanted it to be something MIT-related and personalized, so I went to the campus bookstore and searched around. A book? No, too easy. An MIT shirt? No, he wasn't my son attending school. A shirt with equations on it which, when solved, spelled "MIT"? Tempting, but he did have to make new friends in Chicago after all. Finally, after pacing up and down the two floors for an hour, I saw a picture frame and an idea clicked: I wanted all of Team X to sign it, and I wanted to put in it something that encapsulated our time together. I was hoping it wasn't too awkward because it was one of those ceremonious frames that were meant for two rectangular pictures to fit in, each flanked by a big "&" in the middle.

I chuckled, because _____ & _____ was clearly meant for partners.

Instead, I wanted to put our 2012 paper on artificially reactivating memories in one frame on the left and our 2013 paper on creating false memories on the right. We coauthored our first paper as "Liu & Ramirez," and our second paper as "Ramirez & Liu," so the frame inherently embodied the team-oriented motto we were carrying forward. It was a literal positive memory, framed and ready to be shared. I had all of Team X sign it, and when I gave it to Xu, he tried to feel the gift through the wrapping paper and thought it was a book.

Really, Xu? I got you Principles of Neuroscience *for your frigging going away?*

He opened it and exclaimed, "Wow, this is unbelievable. Did everyone really sign it?"

"Cheers!"

Xu said it would be the first thing he'd hang right above his desk in his swanky new office at Northwestern. He held the frame on his lap for the rest of dinner. There's a moving target we all aim for and sometimes hit; right at the bullseye is what we call happiness. Xu was exactly at its center that evening.

All of Team X remained as the bar began to close down for the night. One by one, everyone hugged Xu, said their goodbyes, and caught rides home. By two in the morning, Xu and I were the only ones left. We looked out at the empty tables in silence, like musicians at Symphony Hall once everyone else has gone home, listening for the reverberating echoes of a memory. I knew that I wouldn't see him for the next few months until we met up at a conference in Toronto. I wasn't ready for our night to end. We had just spent the last five years eating dessert together and calling it science, and now it was time for the check. It is here that the purely "positive" part of this memory ends.

There is a scene in the film *Inside Out* where the character of Sadness touches a golden orb that contains a joyful memory. The once-happy memory begins to radiate with a deep blue hue. It transforms into something deeper, into something more moving, as it's infused with hope and nostalgia. At the end of the night at Green Street, I formed a similar memory that was gilded and cerulean. It was bittersweet—my last night with a knight of science.

The bartender eventually kicked us out, and Xu and I stood on the sidewalk among a serene shower of snowflakes. I was mentally preparing to say goodbye, and I could feel that he was too.

How did it get so late so soon?

We were using any bit of small talk we could think of so that our conversation wouldn't end.

"How are you going to decorate your office?"

"What did you think of tonight?"

"Are you really going to drive straight to Chicago?"

I had the same butterflies I felt before our first lab meeting together, nervous that I wasn't prepared for the moment. My eyes welled up as I thanked Xu for giving me the best five years imaginable in graduate school. I thanked him for being my mentor. I thanked him again for believing in togetherness.

The arcs of our scientific careers were just beginning, and they would always bend toward friendship, we reminded each other. There was simply no way this would be the last time we had a reason to celebrate together.

"So, kind of cliché, but we both know this isn't goodbye, right? This is an 'I'll see you in Toronto,'" Xu remarked.

"Toronto is so soon! And in the summer we have to plan some kind of vacation, and when our paper is out we'll have another excuse to celebrate. Oh, and dinners in Chicago for sure!"

I didn't want to leave anything unsaid or undone so that it wouldn't go unremembered. But I had to stop talking. Each successive word felt like someone was inflating a balloon in my throat.

Xu spoke his mind: "When I leave, let's be okay with needing each other."

On that frosty, moonlit sidewalk, at the end of one last winter evening together, we shared our second hug. The warmth from Xu's arms around my shoulders reminded me that every little and big thing was going to be okay. Somehow, it will always be okay because no matter how far we're apart, our togetherness remains. The last memory I have of Xu is of us hugging on that cold January evening under a quiet Cambridge snow. That memory for me, that wistful engram, has become crystallized as a moment of magic in my mind—a moment that contains all the meaning with all the multitudes that can be found in life.

Part II

6

A Second and Forever

SO FAR, our quest to manipulate a memory has focused on the past. We've found cells that house previously experienced events and have seen how researchers can turn these cells on and off to activate or inhibit memories. We've covered how to create false memories and how to even use memory as a tool to combat disorders of the brain. Still, this is only half of the hourglass of experience. We already know that at any given moment, the future is dissolving into the present, from which the brain builds all its palatial dunes of the past. But in this chapter we'll cover how these dunes themselves are the substance on which the future is built—once flipped, the sand at the bottom bulb of an hourglass inevitability becomes the substance for creating new and practically infinite forms. From the perspective of neuroscience, the sands of memories aren't used only to build our inner worlds of the past but to build every possible sandcastle that we can dream of for tomorrow.

My last year in the lab, I felt like I was counting each final grain of sand squeeze its way through the neck of the hourglass, which was just about ready to flip over once more. My future was incoming at the speed of holy-shit-this-is-stressful—I needed to finish my PhD soon. After five years of graduate school, I was on autopilot. I'd swipe into our lab space, pour myself a cup of coffee, and head over to my workbench, where Henry, my good luck charm in the form of a green blowup dinosaur, was hanging out by the window. Henry, my running shoes, pictures of my family, and a stack of research papers all gave my desk a familiar, comforting feel while I worked. My schedule was so predictable, so consistent, that I sometimes even dreamt of this exact scene.

During my breaks, I often thought about what my future with Xu might look like. While sitting at my desk, I'd mentally travel forward in time and imagine paying him a surprise visit in his new lab in Chicago. I'd imagine Xu standing next to a few students in his lab, showing them how to transfer brain slices from a small plate onto a microscope slide. I'd give him a lucky blowup dinosaur to call his own and make some joke about the cup of coffee I brought him from his lab's own supply. We'd head to his office where his red scarf, beige messenger bag, and pictures of his family would provide him the same familiar and comforting feeling that my personal belongings provided me in my lab.

We project ourselves into the future all the time. When we mentally fast-forward to a possible episode in our lives, our brain patches together a firsthand account of what later may be like. On short timescales, we plan things like lunches, when to meet up with friends and colleagues, and what shows to watch. On longer timescales, we plan vacations, prepare for starting a family, and set goals for that house we hope to buy one day.

Like my imagined future with Xu, these possible worlds are all built from experience—from memory. Neuroscientific research has revealed that when we think about something that hasn't happened yet, we import our related memories to serve as a foundation for imagining how that event may play out. In a series of studies from Kathleen McDermott's lab at Washington University in St. Louis, the researchers asked people to imagine themselves in an environment they were familiar with, compared to an environment they had never visited before. The researchers found that a person's imaginings contained more contextual detail and activated brain regions known to be involved in episodic memory if it was built on the terrain of a familiar memory. This would be like me imagining a visit to Xu's office in Chicago versus imagining visiting Xu at, say, Yellowstone National Park. Though I had never visited Xu's new office, I've been to plenty of offices in my life and could paint a fairly thorough mental image of what the space would look like. But I've never been to a national park, and so I would have a much more difficult time detailing the experience of visiting one. In this way, our memories of the places we've visited in the past can be used to mentally design similar places we haven't been to yet.

Our experience of time matters too. A related study from researchers at the University of Geneva, Switzerland, and the University of Liège, Belgium, found that most people can remember more sensory and contextual details of an episode that happened a few seconds or minutes prior than they can from an episode that happened days, months, or years earlier. This is perhaps intuitive: it's easier to remember exactly what just happened, before our memories fade. Pulling ourselves out of the present and projecting into the future follows a similar rule: we are better able to imagine the details of what will happen to us seconds and minutes from now, compared to years from now. Just as our memory fades, our imagination blurs as it gets further into the future. The closer the before and after are to *right now*, the more detailed they will be in our mind. Like a candle, we cast the brightest light on past and future moments that are closest to the present.

When memories reverberate along the timeline of our past, present, and future selves, they also take information with them. This means that I can't forecast tomorrow without looking at what happened yesterday. Neuroscientists call this the constructive episodic simulation hypothesis: the brain imagines a future by deconstructing its past experiences into experiential elements—sights and sounds, contexts and emotions—and then flexibly recombining these elements into a new simulation of tomorrow.

It follows, then, that manipulating any memory would, by default, change the contents of our imagined future. The brain does this naturally: it manipulates its own memories for the purpose of simulating an unfathomably broad set of possible experiences. It's how we imagine things while we're awake. And because our brain is the ever-active organ that it is, it does something similar while we're sleeping. We've all had a front row seat, while asleep, to the brain creating its simulations; this very process is what is thought to create dreams. Today, neuroscientists are discovering that, in addition to there being a direct link between imagination and memory, there is also a close biological link between imagination and dreams. Like imagination, dreams appear to be integrating the past, present, and future—an organic manipulation of our own memories for our first-person self to see—but in a manner that doesn't require wakefulness to take place.

Why do we dream? This question has been around for centuries. Aristotle proposed that experiences when we are awake are like projectiles set in motion. As these experiences become memories, they continue their travels in our dreams. Two millennia later, Sigmund Freud took a different approach and argued that dreams were the royal road to the unconscious. Whereas Aristotle thought that dreams were caused by the movements of our "sensory organs" and Freud thought that dreams reflected our repressed desires, a more contemporary theory holds that dreams evolved as the brain's way of simulating its own future experiences. These simulations are like dress rehearsals that get as weird as they want on the grand stage that is your brain.

Aristotle and Freud had a point—dreams contain information based in experience—but neuroscientists now propose that dreams are windows into the brain's strategic process of inventing possibilities. And it is memory that allows the brain to recreate an endless number of possible experiences. Whether awake or asleep, the brain never stops remixing its own history. What Xu and I did in the lab to artificially manipulate memories is something that our biology routinely succeeds at doing with and without our conscious awareness.

Moreover, if dreams are truly created from pieces of our past experiences, then sure, they can be pretty bizarre—but they are not at all random. It is in the great sensorium of wakeful experience that we can find the echoes of our dreams and vice versa. This implies that what we call *imagination* while awake may relate to what we call *dreams* while asleep. To exist, both require a latticework of memories.

In fact, the only limitation in our ability to conjure up a future is the expanse of our memory: no matter how hard I try—even if I close my eyes and really, *really* try—I can't imagine a single future scenario with Xu that isn't in some way built from pieces of my previous memories, and the more memories I have, the more mental material I have to work with. Memories are my brain's building blocks. And though this means my dreams and imagination are constrained by my memory—and memory is constrained by both—the combinations and permutations of these blocks are practically infinite. A natural biological property of memory is the ability to manipulate *itself*.

This is because the brain in general and memories in particular are flexible. I can remember the picture frame I gave Xu at Green Street, pull it out from my memory, and hang it up in an office space I imagined for him. Or I can mentally manipulate this scenario and bring my dinosaur, Henry, to Xu's office and place the picture frame in his T-Rex arms for all to see. Or if I'm feeling particularly imaginative with my memories, I can transport all of Team X to Chicago to hold a celebratory ribbon for Xu to cut with oversized scissors to signal the grand opening of his lab. We all have memories of the past and, in a very literal biological sense, we all have memories of the future as well.

Our brain produces a sort of mental entanglement between past and future events where one cannot be described independently of the other. Even so, we should be able to disentangle them in the lab. In other words, if the brain really uses memory to sculpt the future, then we should be able to look at neural activity from yesterday to predict how brain cells and, therefore, an organism will behave tomorrow, and vice versa. In the book *1984*, George Orwell wrote, "Who controls the past controls the future: who controls the present controls the past." It turns out that this holds true in neuroscience, well outside the realm of fiction. And I would now extend this to: "Who controls the *future* controls the past." The *who* here is, of course, our brain. The rest of this chapter is about how memory is the very thing that fills our dreams and fuels our imagination so that we can move forward in life and make even more memories.

———

We've seen in previous chapters how damage to areas like the hippocampus produces amnesia and how direct stimulation of this area can enhance or restore memories, but what does the hippocampus *do* when processing experience in real-time? How does the brain create a memory that links past, present, and future?

These questions have been around for decades. Nobel Prize–winning work in the 1970s revealed that hippocampus cells tend to fire when an animal visits particular locations of an enclosed environment. When an animal is exploring a maze, for instance, one cell will fire when the animal is in the upper right corner of the maze but not the other corners; another cell will fire in the lower left corner and not the others; another cell will fire in the middle of the environment and less so up against the walls. Many believe that these cells are collectively firing in a sequence to form an internal map of an environment.

These cells are aptly called "place cells," and they have been detected in both rodents and humans.[1] As we saw in chapter 1, the hippocampus generates patterns of cellular activity when an organism is forming a memory. The brain then begins rehearsing what might come next, as if it's ruminating on recent events: the sequences of cellular activity that were present while forming a memory are prominently *replayed* when the organism is resting soon after. It's biologically advantageous, presumably, to internally prepare for a future that is uncertain.

But there's an interesting wrinkle in the process of memory formation. Over the last decade, and well after the discovery of place cells, neuroscientists discovered that the sequences of neural activity we observe when an animal is making a memory can be detected even *before* the memory is formed. The hippocampus *preplays* the same activity patterns that will then reemerge as a memory is formed.

The brain is in fact constantly predicting what will happen next. And it is rehearsing *multiple* outcomes because, if it can control the one it will end up in, it will be more likely to survive into the future. It's as if our brain is constantly rephrasing, "What are we going to do once we get there?" to "What *can* we do once we get there?"

And once *there* arrives—getting to work, a friend's apartment, a wedding—some of the same patterns of cellular activity that were preplayed in anticipation of the event are then replayed right after the experience actually unfolds. Thought of this way, preplay is a template for subsequent experience to build on. It's the brain's way of arriving to *now* with adequate preparation. Once a memory is formed, the hippocampus then *replays* its contents to consolidate it.

The relationship between preplay and replay is analogous to a digital piano. Often, these pianos come preprogrammed with song templates, which we learn to play by pressing keys that light up in a sequence to indicate the order to play them in. We rely on the preexisting sequences, which are part of the piano's standard hardware, to successfully guide our experience of playing them.

The brain's preplay events thus reflect a preexisting biological template, on which experience is built. Neuroscientists call this *prospective coding*, which is when brain cells fire in sequences that represent possible future scenarios. Experience, then, iterates on these sequences and strings them together into something new, into some unheard combination that has structure to it. All brains, therefore, come preprogrammed with keys ready to become music.

These playable sequences of memories in the brain carry usable information. Rats navigating a maze show cellular activity in a manner that predicts an upcoming choice to go left or right, for instance. As neuroscientists have noted, this cellular activity is observed even when the animal isn't performing any task or showing any obvious changes in behavior, such as while they are resting or sleeping. In humans, the constructive episodic simulation hypothesis holds that we actively recall memories in order to extract the information needed to build our simulations of the future. Since these sequences can be bound together in new combinations, they are like the brain's physiological correlates of simulating upcoming events. Neuroscientists have reasoned that an organism with this capability can then make an adaptive decision and plan a nearly infinite number of scenarios accordingly.

In trailblazing work from 2024, researchers at Harvard Medical School studied the brains of mice as they formed memories and found that the hippocampus and cortex contain cellular activity patterns that can be used to predict what an animal will do in the future. Similarly, when that moment arrives, the cellular activity patterns that emerge can be used to decode what the animal was doing days beforehand (and the contents of its memory from that time). A group from New York University even found that the hippocampus displays unique patterns of activity during select moments of wakeful experience that are replayed numerous times during sleep to promote the consolidation of those moments into memory. These patterns, too, could be used to decode exactly what the animal was doing at any given time. The brain's preconfiguration anticipates tomorrow by default and uses experience to reconfigure itself over time.

It's amazing to me that the patterns of neural activity that the brain preplays before a memory is even formed are also observed during sleep. Neuroscientists believe that sleep is when recently experienced memories are consolidated in the brain for long-term use—but this is only half of the story. The other half is that the function of sleep may be to let memories blend—a process which we experience while dreaming—into something new, like new solutions for problems that haven't yet happened. Sleeping, in general, and dreaming, in particular, may act as the brain's training facility where it can learn from the past and let itself improvise along the way. Dream states are some of the most free-form states a brain can be in, as our memories are given a chance to braid themselves together into new experiential prospects.

In an influential study, psychologists found that people were twice as likely to figure out a hidden rule in a math-based task if they slept shortly after training. The researchers proposed that sleep is when memories are remodeled and qualitatively restructured in a manner that enables insight. Dreaming perhaps gives us a first-person account of this reorganization happening.

So far in this chapter, we've seen that neuroscience is beginning to solve the riddle of what I'll call the mental wormhole: our brain's ability to bridge any two moments, regardless of whether they've even occurred yet. We're seeing that the act of recalling a memory and the act of imagining ourselves in the future generate similar patterns of activity strewn across similar brain regions. The human hippocampus is no exception: subjects learning to navigate a college campus showed a similar pattern of brain activity as when they imagined walking through the same route. The researchers could even use the brain's physiological activity alone to decode and reconstruct the real-world and imagined positions. As other researchers note, not only do memory and imagination tap into similar neural processes, it turns out that damage to areas like the hippocampus dramatically impairs our ability to use imagination, to plan out future scenarios, and even to dream. Remembering, therefore, may be the brain's way of imagining the past.

This all means that there is no imagination without memory and no memory without imagination. When it comes to how the brain stores and uses information, we're looking at a much deeper biological principle at play, which extends beyond our concepts of space-time. The brain's physical relics of the past, of the engrams of what once were, are its projections of what hasn't happened yet. And the brain's projections of the future, of the engrams that will be, are also its projections of what has already happened.

It's worth noting that while imagination lets us mentally roam freely through our world of memories, it can quickly become a tool to use in a moment's notice. Imagination can have a very practical purpose, but when it becomes less about personal experiences and more about combining bits of knowledge in new ways, we get *creativity*. I think of creativity as imagination's more immediate, solution-based counterpart. Unlike creativity, however, imagination can be as unconstrained as we want and doesn't necessarily require an immediate application in the real world. The two are psychologically and neuroscientifically related—both invent something new and can even complement each other, and both depend on the hippocampus. Scientists have found that our creativity is enhanced if we first recall details from a recently formed memory, as if the memory itself is revving our brain's creative engine.

Imagination, on the other hand, doesn't have to be useful in the moment. I can imagine walking into Xu's office or going on vacation with him to Cape Cod just because I want to lose myself in such thoughts. Creativity lets us generate new ideas and ways of thinking by melding together information in a manner that doesn't always require using the episodes of our past. What if you dropped your keys somewhere on your way home from work? You could use your memory to retrace your steps and imagine where they may have fallen, but to find a solution you'd need to use your creativity: Do you try to break into your home by bending a paperclip a few different ways to pick the lock?

Or do you call the building manager at work to ask if anyone has seen a pair of keys on the floor? Or do you buy a metal detector and go searching for brass?

Of course, some creative solutions are more effective than others, but the point stands that creativity consists of combining knowledge in new and often unusual ways, while imagination involves combining memories to simulate future moments that we may experience. When someone prompts you to "Be creative," or to "Use your imagination," the former is often to solve some problem at hand, while the latter serves to entertain multiple possible outcomes. Both are cognitive tools that the brain calls on to think outside the box.

If memory is constantly engaged to predict the future, how does the brain react to a future it had settled on but suddenly is no longer possible? If a flight is canceled or a friend texts that they can't make it to dinner, everything has to shift. The brain comes equipped with cells that signal when the predictions of future events need to change. These cells respond immediately when reality deviates from what was anticipated, leading to a so-called *prediction error*, so that we can make more accurate predictions in the future. One speculation is that these cells are teaching the brain how to learn: if an outcome occurs exactly as predicted, such as walking into the lab and seeing Xu by his desk, then I don't need to learn much. Reality matched what I had anticipated would happen. However, if I walk into the lab and see Xu sitting at *my* desk, I know something must be up—we've either switched desks, he's playing a joke on me, or I'm mistaken about where I sit. In all cases, I'd have to learn why things didn't line up with my expectations.

The brain is really good at picking up on these changes because it is constantly sampling our experience for any inconsistencies that deviate from what it expected. Perhaps the most powerful inconsistency I know of is when we think we're going to see someone—someone who, in fact, we will never see again.

How the brain adjusts to the loss of a person is a process without an obvious resolution. When an engram of a friend no longer has the opportunity to play itself out in the world, when the chance to meet them again becomes permanently confined to the stuff of imagination and dreams, the edifice of memory comes crashing down, and we are forced to confront a reality that we've never thought was possible.

————

I woke up Thursday morning, February 12, 2015, to the tickling feeling of my phone vibrating under my back. I had fallen asleep on it at some point the night before. Out of habit, I reached for my sunglasses to dim the morning light in my room, as brightness always strains my eyes and leads to a headache. I went through my usual routine of checking out social media and then scrolled through my inbox to see if there were any time-sensitive emails I needed to respond to. I saw a subject line that stood out. "I hope this isn't true," it read, and the email was from a professor at MIT.

I took my sunglasses off to read more carefully. I read what turned out to be a brief series of emails between a handful of professors in Boston, California, and Chicago.

"What happened?"

"Are you sure?"

"Has this been confirmed?"

"Who else knows?"

I didn't know what to make of it and then I read:

"How did he die?"

I looked away from the email, knowing that what I was about to read could only get worse. I let out a labored, "Fuck," and sat up in bed. My first thought was that a senior researcher in our field had passed away and that these professors were emailing back and forth in disbelief, as happens when death interrupts all that was planned. I continued reading.

"My friend told me that they found Xu at his home." I read the sentence again.

"My friend told me that they found Xu at his home."

I panicked. I called him. No response. I called again, ready to apologize for calling so early in the morning since he was an hour behind in Chicago. No response.

He's probably sleeping, I reasoned. And if they had "found" him at home, surely he was in the hospital recovering or under intensive care and in good hands? I googled his name and searched his name on Twitter, but nothing new came up.

I rushed to text Xu again: "Please tell me you're okay?"

It was barely seven in the morning when I received a personal email from Susumu. "I'm so sorry," read the subject line. I didn't have to open his email to know that I had lost Xu. I buried my face in my pillow and quietly started to cry. I felt like all my limbs were desiccating and meeting at my chest, emptied of any hope.

I shifted to questioning the accuracy of the information. How could a healthy thirty-seven-year-old die all of a sudden? *How* did he die? Why?

I couldn't make it to my bedroom door without breaking down. So I sat in my bed and just stared at the wall in disbelief, my eyes squinting and strained with tears, each painful glimmer of sun forcing me to dig my fingers deeper into my palms, scratching off pink, throbbing skin in my panic. We hadn't messaged in a few days, and I kept checking my texts to see if the blue dots that indicated Xu was typing would pop up.

I finally left my apartment for the lab, the only place that truly felt safe, the only place that made sense.

I called my dad on the way without really knowing what to say to him. He answered with his cheerful routine: "*Hola, buenos dias, hijo! Como has amanecido?*"

I didn't respond. He said hello again, and I barely got out the words: "Dad, do you have a second?"

His voice lowered. "Mijo, for you, I have a second and forever."

I couldn't speak. I hung up and found a frosted park bench nearby to sit down, as I knew he would call back. My dad's words cut me to the core. Xu and I talked about memory that way: it begins within a few seconds, it enduringly modifies the brain, and it can last forever.

I picked up after two missed calls and told my dad what had happened. He told me he loved me and that this kind of pain would hurt for the rest of my life but that it would change me in ways that imbue my experience with an added layer of purpose. Without intending to, my dad had hit on one of the deepest, undeniable parallels between grief and memory: both endure across the entire lifespan, forever changing us, helping us to decide what matters most.

When I walked into the lab, a handful of Team X members were in the coffee room, puffy-eyed and dejected.

"I'm so sorry, Steve," someone said, and I elbowed the door behind me and ran out of the room.

One of the Tonegawa Lab members found me in the bathroom, sitting on the ground against the wall. She sat next to me and put her head on my shoulder.

"How the fuck does this happen?" I asked, out of breath.

"I don't know," she said, "but I do know what he meant to you. Don't feel like you have to make sense of it today."

In search of some kind of closure in the blur of hours that followed, I kept asking myself, "How did he die?" And the single memory I have about the details surrounding his death came from a heartfelt conversation with his girlfriend, Jessie.

"The night before he died, Xu called me and talked about how he was excited to have already made some new friends in Chicago," she said.

We both knew that this meant a lot to him. We paused the conversation to gather ourselves. Jessie then told me that the next day, Xu's colleagues had become concerned about him after he had uncharacteristically missed a few meetings at Northwestern. They asked the police to check up on him, in case anything had happened, and they found him shortly after in his apartment. He had passed on February 8. The details surrounding his passing would remain private because some memories are so heart-wrenching and so unimaginable that they transcend the human urge to be shared; they cannot live anywhere else but home, where they can be protected, mourned, and absorbed. Without the details, closure would look different for everyone, and that's okay; for me, whatever the reason for his death, the outcome was still the same: my friend wasn't here anymore.

What is the last thing Xu thought?

In the last moments of his life, Xu's brain formed a memory never to be remembered—one in which consciousness is fighting to keep its eyes open, while warm biology transitions back to the cold touch of matter. I think about how this final elusive engram briefly existed as unique patterns of brain activity in motion, and now lastingly rests as its component molecules take on new, breathless forms. The substances of life morph just as memories do; the fabric of reality is constantly changing. And I think about my hope—my deep, unwavering hope—that in his transfiguration between matter and memory, Xu found harmony. We are all guaranteed to have this one final stamp of experience leave its mark on our brains, an engram cut short that can never be shared.

My engrams of Xu are perennial, but the force they came with during the first weeks and months broke me. Imagining his death reawakened the kinds of memories that I tried not to think about because they left me feeling completely helpless: the excruciating roar from Xu's father as he willed himself to speak in agony at the funeral, or the empty look on the faces of his family members as they walked to their seats in the chapel. The thought of death and all the ways in which it can arrive consumed my imagination in ways I did not understand.

———

As hard as it is today, thinking about someone when they're no longer with us has brought me a breathtaking sense of wonder for what memories can do. When my grandmother died, my dad taught me that those we lose rest in our mental memorial permanently. "When you remember someone, you reconnect with them," he told me. It turns out that there is a neuroscientific basis for this reconnection.

The first reconnection with someone we have lost is the hardest because it's the most unfamiliar. Although I didn't know it at the time, the moment I began to reconnect with Xu was right after I sent him the text that would go unread forever—"Please tell me you're okay?"—which is when I was no longer talking to him but to the memory he had become.

Imagination uses memory to structure itself with substance. It's the brain preparing to turn its own activity into some external influence on the world. The way our brain imagines the future parallels the way it remembers the past. But when the substance of imagination goes unused—when it *must* go unused because the object of the memory is no longer with us—the brain is forced to subtract these possible worlds from its list of future scenarios. It has to make peace with the brutal reality that some of its plans will forever remain confined unto itself. What was once the brain's precomposed and certain future is now an unquiet loop of neural activity suspended in a mental limbo. A memory's search for meaning, for being actualized in the world, begins anew. The past wants to be learned from again to construct a different future. All the memories I had of Xu that were building one future together had to be repurposed. This is where grief begins.

Grief is the natural response to loss. It has biological purpose, researchers argue, as we begin to make sense of a world devoid of someone who had always been in it. It may even be an adaptive form of learning. Like the making of memories, grieving needs time to fully develop in the brain. And as grieving occurs, it holds the power to occupy our imagination and dreams. It is persistent and leads to insights that underscore just how powerful the memory of a lost friend really is.

The moment we find out that someone has passed away, this information flies in the face of the thousands of interactions we've had with them. Our worlds of experience with that person all have to zero in on the single fact that they're no longer with us. Cognitive psychology teaches us that while the knowledge of someone's death is immediate, how we incorporate it into our understanding of the world is not. For instance, upon hearing about someone's death, our brain has every reason to make the prediction that the person is still alive since we've had so many more moments *with* them alive than *without*. The brain determines this is the most likely outcome but creates an emotional dissonance in the process between expectation and reality.

The brain even goes out of its way to make sure it physically holds onto memories of people that matter. In mammals, for instance, more brain cells become active when a social bond is made compared to when a nonsocial event is experienced. This increase in neural activity can even predict the strength of the bond itself. Visiting with the same social partner means the brain will have direct access to the cells holding onto this social experience and stimulate activity in these cells. A somber prediction is that a partner's passing will now leave this *exact* pattern of cellular activity quiet for the rest of the organism's life. The cells that breathe life into the memories of a partner continue to exist, even when that partner does not. They are perhaps only partially reengaged when imagining what could have been or dreaming what we wish were so.

This poses a new biological challenge. The brain routinely makes the prediction that others in the world are persisting even when we're not physically with them, but the brain cells holding onto these memories are only fully engaged by the physical presence of the individual they're holding onto. When that individual passes away, our internal model of the world breaks down and needs to incorporate new information in the form of someone's permanent absence—perhaps allowing the brain to form new connections with the cellular representation of that individual and broadening both how we access related memories and what they mean to us. The engram of a friend becomes a ghost in the brain, walking only through the mental corridors that our imagination and dreams have built.

Integrating this reality into our mental expectations of the world takes time. Initially, the brain delivers jolts of stress hormones when a bond is broken in some way, such as when someone goes missing. This violation of our expectations puts our brain into this search mode as we hunt for some kind of answer—texting, calling, looking for them on social media—all the things I did the morning I found out about Xu. But just as memories are updated, little by little, each time we recall them, it takes multiple experiences with someone's *absence* to fully accept the reality of their passing.

From the perspective of the brain, a similar mechanism may be at play not only when someone close to us dies but when we lose a connection to someone very close to us—for instance, a partner ending a relationship, a friend severing ties. In each case, we grieve the loss of a future we presumed to be certain. In each case, a core emotional network of the brain is engaged, sometimes to the point of chronic dysregulation, which can lead to prolonged bouts of depression. The feelings that these lost connections conjure up extend to the inanimate too: our home may be lost in a storm, we may lose our job, or a piece of family memorabilia may be stolen. Despite differences in intensity, quality, and duration, when we experience loss, the brain has to reimagine a future by resetting its expectations. This reset, not surprisingly, is grueling.

I believe that loss happens when all of the building blocks of memory topple over—when the very foundation that we used to imagine tomorrow and to dream of a future suddenly collapses. If this is true, then I also believe that the reverse process of recombining these building blocks, one by one, into yet another future is where healing begins. This phase perhaps engages a rival pattern of activity in the brain's emotional networks that begin to rewrite what the loss means. I wonder if this phase is why dreams are so vivid after such emotional extremes: the brain is attempting to build a brighter future for itself.

For the time being, the reality for me was that I depended on Xu. I depended on him for feeling secure in the lab and feeling confident as a researcher. As long as I had him as my lifelong friend in science, I felt fearless in what we could accomplish. But I was scared now. My belief that I may not be competent enough on my own as a scientist stemmed from how much I had depended on him as an ever-present lab partner. Without Xu, and with this dependence completely severed, my life in science and role as a scientist were obscured by uncertainty. I went through feelings similar to those felt when losing a spouse: life became emptier, my sense of identity was diminished, and it became nearly impossible to imagine a new future. It's as if the brain's way of grieving is to first become lost. The brain shakes up its own foundations to the core—down to its very beliefs of how the world operates—before it can then begin to reorganize all that it knows and expects.

———

I walked into the lab and was ready to start working. I swiped into our space, poured myself a cup of coffee, checked around to see who was in our lunchroom, and something felt off. Did someone move our tables?

Everything was hushed. I blinked and found myself at my desk checking to see if it was organized the same way as I had left it.

Henry, my green, blowup dinosaur, was still there by the window. My running shoes were still where I'd left them, tucked beneath my desk. There were research papers stacked in a pile, and my pinned-up pictures of family were exactly where they should be, right in front of my computer monitor. A handful of people I didn't recognize were walking through the lab—I couldn't make out what they were saying, as they sounded monotonous and subdued—and when I tried to say hello, my mouth felt like it was full of peanut butter. I just mumbled and stopped speaking. They didn't notice me and kept walking. I tried to follow them, but my legs were weighed down.

I looked around some more and saw a thin man wearing dark blue jeans and a red scarf neatly tied over his red flannel shirt. He was hunched over a desk looking for something, a pen perhaps. For some reason, I couldn't stop staring at him, and when he turned around, my eyes widened and I fell to the ground, leveled and feeling like someone had dropped a dumbbell on my chest.

"Who else knows you're around?" I asked.

"It's going to be okay."

"Let me go with you instead, please," I begged. "Please."

He didn't respond. Everything around me couldn't possibly be true. I sprung up and jumped, levitating a few inches off the ground for a second longer than I should be able to. I tried to pinch myself, but my skin felt like the rubber of a worn-out tire. I forced the tip of my index finger underneath my thumbnail, but nothing registered. I dug deeper, blinked, and my index finger turned into a toothpick. I pulled back, tensed my entire body, and then dug my finger back underneath my thumbnail, which it readily absorbed without registering pain.

When I looked up, Xu was gone. I knew what I was experiencing wasn't real, and as I tried to run through the lab to find him, my arms and legs felt like they had resistance bands pulling them backwards. I broke free and looked around the lab, which was now full of rain clouds.

This had to end. I teared up and sprinted headfirst toward the window by my desk, shattering it as I plunged forward toward the Stata

Center in front of our building. The five stories below me transformed into the park that was next to my childhood home.

Just before I hit the ground, I woke up in a cold sweat; my feet felt like they had been submerged in an ice bucket. I turned on my phone's flashlight and quickly realized I was back in my bedroom in Cambridge. I knew I had just experienced a lucid dream.

A lucid dream is when you become aware that you're dreaming and can begin to willingly warp your fantasy-like experiences in your sleep. I used to lucid dream nearly every night in high school and college, and then that ability went away in graduate school. Shortly after Xu passed, though, I began lucid dreaming again, and my dreams were more vivid than I ever experienced.

I'd randomly wake up in the dream itself, at first shaken by the surreal. There was little comfort in its ghostly visions. These dreams were often the same: I "woke up" either in the lab or somewhere abroad and had no idea how I got there—I usually don't just wake up all of a sudden halfway across the world. I became aware that I was dreaming after a simple test: I'd pinch myself by spearing my index finger deep underneath my thumbnail to see if I felt pain. The second it felt like rubber, I knew I was still asleep, and I had what felt like a few minutes to explore my own shape-shifting thoughts.

Everything in this world was built from memory—floral carpeted hallways from hotels I had stayed at would lead to sun-kissed rooftops from restaurants I had eaten in would lead to neon-lit arcades I used to frequent. If I fell asleep listening to music or to a podcast, I could still hear the contents during lucidity, and sometimes they would even make their way into my dreams, reorganizing the world around me and filling it in with its spoken details, often sounding like people speaking underwater. During my short-lived suspensions of reality, I'd frantically search for Xu, and once in a while, I'd find him. He usually had gray hair and appeared to be in his fifties, his eyes made heavy by an unspoken sadness and his face lined with wrinkles. Each time I found him, he said he needed to go away—he never told me why but always said he'd be back soon.

I couldn't avoid the psychological aftermath of losing Xu every evening. My mornings began by awakening from a place where I had already been "awake." It was like being on an eternal commute, with each leg of the trip leading me further away from a place of rest. When I woke up, I'd feel the same emptiness I felt when I'd read the email about Xu's death. It's the kind of hollowness that makes you question if you're even *living* because experiences without feelings don't feel like life at all.

———

Recurring thoughts of my past with Xu and my future without him consumed me. I began lucid dreaming after his death in 2015, which was unavoidable given how much I thought about him in my waking life. But I wasn't okay with sheer speculation when it came to the reasons for memory's ability to populate dreams. I wanted to know *why* my brain granted me volitional access to the worlds built in my dreams. Why did my brain unlock every possible reality while I slept and then place me right in the fantastical middle of unbounded subconscious? Words etched on the Temple of Apollo tells us to "Know thyself," but it makes no distinction between whether thyself is asleep or awake, or both. I'd like to offer a modern take on this maxim: Know thy memory. It contains our conscious and subconscious self—it contains *all* of us.

Lucid dreaming has direct access to our past and to our future. Researchers have hypothesized that, like the reconstructive nature of memories and false memories that I explored in chapter 4, our capacity to restructure our past during a dream state may also enable our ability to *imagine* the future. Compared to patients with intact hippocampi, patients with hippocampus damage report dreams with less detail and also show impairments in projecting themselves into the future. In the undamaged brain, our ability to mentally time-travel between episodes of our lives, "could provide the neural building blocks for simulating upcoming events during decision-making, planning, and when imagining novel scenarios," as neuroscientist Randy Buckner proposes in his 2010 paper. In humans and in rodents, the hippocampus is active even when the mind is idling as well as when it's dreaming, which may explain why my lucid dreams were so often about Xu: my brain is using our previous experiences together and imagining the future, *any future*, it so desperately wants with him. If this is true—if this is in fact the scientific reason why my will is free to move during moments of lucidity—then I believe one of the most amazing human faculties is enabling my lucid dreams: hope.

Hope is using memory to imagine a better future. The brain does this so effortlessly when we are asleep. Studies conducted in 2012 and 2018 that sought to understand the biological underpinnings of where past and future meet in dreams found that human subjects undergoing fMRI scans during periods of lucid dreaming have increases in communication between multiple brain regions associated with memory and imagination, including the prefrontal cortex and hippocampus. The activity in these areas even correlated with the amount of content recalled from each dream upon waking up, as well as how vivid the dreams felt. What's more, studies find that nearly half of all dreams are reported to contain at least some aspect of what a person experienced when awake, and this moment in time, such as talking to Xu by his lab bench, inserts itself as part of a larger and fragmented dream scene where imagination is the cinematographer and memory is the director. The brain devotes a tremendous amount of energy to using our past experience to anticipate, *to hope for*, a better future.

In 2021, a group of scientists from around the world went a step further and found that they could communicate with people while they were in the middle of a lucid dream: in their paper titled "Real-Time Dialogue between Experimenters and Dreamers during REM Sleep," they state that the dreamer could both "perceive and answer an experimenter's questions, allowing for real-time communication about a dream." These questions included math problems and yes-no inquiries, and the dreamer answered with voluntary eye movements and facial muscle twitches, amazingly creating a two-way bridge of communication between the scientist and the dreamer. One of the lucid dreamers explained the "bridge" in this way: "I was at a party with friends. Your voice was coming from the outside, just like a narrator of a movie."

These studies underscore the intimate relationship between sleep and memories in the human brain. It sounds farfetched but it's true:

dreams can be interrogated by an awake observer, and the view from within the dream can be reported back. I think this may be why I can feel and hear my body breathing while lucid dreaming; certain kinds of external information can make their way into our dreams, which can be reported back through subtle muscle movements.

As it turns out, some recent research supports my self-remedying method of snapping awake from a lucid dream. A study in 2024 found that lucid dreamers could reliably will themselves awake through a variety of strategies. These included closing their eyes in the dream to "get out" of that state; actually saying "wake up" while in the lucid dream, as if commanding the semi-conscious state to become fully awake; and visually fixating on a single point in the dreamscape to gain a sense of control over their body and its relationship to the evanescent surroundings. All of these methods offer volitional escape from the realm of the subconscious. A potentially therapeutic angle here involves training lucid dreamers to gain reliable control over the contents and actions of their dreams. If successful, one can switch from avoiding lucidity to reorganizing the dream's contents in real-time into something more adaptable. This would mean directly confronting our dreams, our *selves*, and finding a serene resolution.

Even dreaming without lucidity helps us stabilize emotional memories by taking advantage of their malleable nature. Dreams may rebuild memories by associating disconnected experiences and bringing them together to form a new possibility.[2] As one paper in *Frontiers in Psychology* suggests, "The insertion of bizarre items besides traumatic memories might be functional to 'impoverish' the negative charge of the experiences." It's in our dreams that hope may be restored, even in the slightest.

———

After we found out that Xu had passed, Team X and I gathered that evening across the street from the lab to support each other and sit with the disbelief, but at least to do so together. Just before heading out, I received an email from neuroscientist Paul Frankland asking if I wanted to speak at an upcoming neuroscience conference in Toronto—the one Xu and I were supposed to meet up at—using the time Xu was allotted, to reflect on him as a person and scientist. This was the only message I received that day that didn't make me feel lost. I don't know if I was feeling an overwhelming sense of honor or obligation, or both, but the request gave me a sense of *purpose* to carry our engram torch forward.

I'm so lucky to have experienced the papers, talks, and awards with Xu, the pageantry of it all. And yet the only thing I could think about was wishing he was with me to share a third hug. Now I could only do that in my dreams.

I remember once telling Xu, "It kicks ass to do science with you."

He responded, "How *else* would we do it?" as we clinked our drinks together.

Now, for the first time in my career as a scientist, I was forced to figure out how *else* to do it.

I used to imagine walking through life side by side with Xu. I'd dream of celebrating our discoveries together and watching as our labs grew into fully functioning teams, as we continued our search for the inner workings of an engram. How lively these memories once were.

The idea of an engram is that experiences leave measurable changes within the brain—it's how we internalize reality. How we choose to externalize our reality is the measure of a person. It's how we leave a real mark on the world, through an engram untethered from a brain, free to influence everything it touches.

It's the reason that I truly believe I still have a future with Xu—one that now lives in my memory.

7

You Are What You Remember

CHANGE IS a concept woven into the very fabric of the cosmos. What began as hydrogen atoms 13.8 billion years ago turned into stars and galaxies, which then turned into all the elements supporting life, until some *became* life. It's nature's grandest experiment: take the universe and change it into biology. And nothing in biology has ever truly stayed still: from molecular reactions to dynamic neural networks, the biochemistry set that is your being is constantly experimenting on itself to permit the persistence of life. Biology tosses and turns, it flutters and soars, it shrieks and sings, it does everything needed to move forward. As we approach the end of our memory manipulation journey, it's at last time that we embrace change as a fundamental feature of the brain and see just how important change is in allowing us to adapt to a constantly changing world as well.

One of the overarching themes of this book is that memories thread and unify our overall sense of being. And yet, our inner biology is a world of unrest. It is built from unstable components, from cells that are constantly recycling their own machinery, replacing the old with the new. So how does stability exist in the form of lifelong memories?

Neuroscience today is teaching us that just because memories *feel* stable doesn't mean that they have to *look* stable in the brain. Yes, they exist in the brain in a manner that can last a lifetime, but the cellular machinery that makes a memory possible today does not have to look anything like the cellular machinery that makes it possible a year from now. It's the brain's grandest experiment: take a moment and change it into an engram.

In doing so, memory gives our autobiographical account of the world cohesion. It gives us a continuous sense of *identity*—that essence of who we are that helps us feel stable. It's our core. Somehow, our stable sense of being is produced by inherently dynamic biological components: synapses grow and retract, the firing patterns of brain cells adjust, DNA gets put to work, and entire systems of the brain strengthen and weaken their communication with each other, all as a result of the never-ending flow of experience. Change is such a part of our existence that having a stable identity seems physically impossible.

And yet, you and I both persist.

In this penultimate chapter, I'll take a candid look at the biology of change and how it enables all the memories and experiences that embody the human condition. I embrace change as a fundamental feature of our biology, as the brain is constantly changing in its activity every single moment of every single day, from the second you take your first breath to the moment you exhale for the last time. Our memories change, and what we imagine and dream about changes too. As a result, *we* change.

I personally find this rather cathartic. While my life changed the morning of February 12, 2015, it's not that Xu wasn't with me anymore. It's that he was now a *part* of me, a life converted into a memory. Perhaps part of me stayed with him too.

———

Once in a while I go back to my family's photo albums to revisit moments of my past. Some pictures immediately bring to life the vivid memories embedded in them. There is a photo of my dad and me at his surprise sixtieth birthday celebration, and whenever I see it, I instantly recall my dad's bewilderment as over a hundred family members yelled, "*Feliz cumpleaños!*" when he walked in. I was twenty-four years old at the time. The room filled with applause and laughter, and some of my aunts and uncles were a bit teary-eyed. My brother, sister, and I looked at each other with excitement and relief, knowing we had pulled off the surprise. It's a memory that feels like it has maintained its emotional impact and episodic detail for over a decade.

Other memories are blurry, almost dream-like, and the further back into my own past I go as I flip through the photo album, the trickier it is to reexperience such moments. As years turn to decades, I feel a strange sense of detachment emerge. In one picture, I'm holding up an album of Pokémon cards and looking into the camera with a real sense of pride. I was fourteen years old at the time. I loved collecting these cards, and as soon as I'd open up a package, I'd put every individual card in its own sleeve and then in a binder full of pocket page protectors to keep them nice and safe. Every week or so, my dad would drive me to the local card shop, and we'd pick out a pack to open together and file away neatly in my binder. I'd teach him about each Pokémon's strengths and weaknesses on the ride home. One afternoon, I scored big and found a holographic Charizard in the deck—this was the holy grail of Pokémon cards. And so, when my mom saw my uncontrollable exhilaration, she asked me to pose with my collection. *Flash.*

Even though some of the sensory details and emotions linked to this memory are still somewhat palpable, I've thought about this moment so many times that I can't help but be acutely aware of just how much of it I no longer *actually* remember. Sometimes I feel like the trace of this moment in my brain exists as a faded outline of the past. It's been about two decades since my mom took that photo, and I can't relive those few seconds of holding up my album the way I could in the days or even years after it was taken. Even though I can still reexperience the emotions surrounding the photo, this treasured memory has lost its episodic luster.

Then there are experiences I've had that are out of my reach—and I get an uneasy feeling of being disconnected from my own past. There is one photo that was taken on my thirty-third birthday—I was at Capital Grille with my parents, brother, sister, and partner, and they're all forcing a smile back at the camera. Everyone (except for me) was uneasy because I had downed one too many martinis on an empty stomach, and I had come close to falling asleep multiple times at the dinner table. In the picture, I'm staring emptily past the camera and with no obvious point of focus. What's more, it was the first time I was ever this intoxicated around my family. If it weren't for the photo, I'd have zero idea what happened that evening. The slice of coconut cream pie on the plate in front of me, my favorite maroon sweater, my mom's hand on my shoulder—all are details that I can see in the photo but can't mentally relive. The memory itself is completely devoid of emotional and episodic vigor.

I know that the young adult in the photo with my dad is me. I know that the teen holding up an album of Pokémon cards is also me. And I know that the semiconscious adult at Capital Grille is me as well. But so much of my life has happened between each one that they all feel like different versions of "Steve" engaging with reality in very different ways. Some ways have led to stable memories, some have led to total amnesia, and some have led to something in between.

If I can't remember all of the details of my life, especially some of the most influential moments that have brought everything from gratitude to shame, then how can I truly know who I am today? My memories sculpt my sense of being, but how does the brain do this when the past can be so unstable, when bits and pieces of some memories feel just as new as the day they were formed, while others feel like they might as well belong to someone else?

These fascinating, albeit thorny, questions about memory and identity are at the forefront of neuroscientific research today, and we have the fields of ancient philosophy and physics to thank for setting these questions in motion. Indeed, the idea that the self is a dynamic thing has been around for millennia.

In Greek mythology, the Ship of Theseus is a thought experiment in which a ship has all its components replaced, one by one, over the years. At some point, none of the original pieces remain, so is the ship still the same Ship of Theseus? To continue the thought experiment, if we were now to take all of the old pieces of wood from the old ship that we've gathered over time and then build a new ship with them, is *this* now the Ship of Theseus? A philosophical approach to memory and identity might say that each incremental change represents the transformations we undergo across a lifetime. As long as there's a continuous history that ties together the ship's structure and function over time, then some semblance of sameness remains.

Physics gives us a powerful insight to work with: change is woven into the fabric of the universe. The atoms produced in the Big Bang forged themselves into stars, which exploded and released their chemicals to form nature's ultimate experiment. This billion-year conversion of energy from one form to another made *you* along the way and everything else in the known universe. We are all recycled star-stuff, in other words, and the recycling never stops. In 1953, a study conducted at Oak Ridge National Laboratory in Oakridge, Tennessee, claimed that 98 percent of all the atoms in a person's body change out every year. While that number is a very rough estimate and almost certainly depends on which types of atoms we're talking about, the notion that our body's molecular components are replaced over time is a basic property of biology.[1] With the average human lifespan hovering around eighty years, this means we undergo several great transformations in a lifetime.

From the perspective of physics, there are multiple different versions of us lurking in the past. If we studied identity with our physicist's cap on, we'd find that our seemingly stable sense of being is made out of atomically unstable pieces. We are constantly changing and morphing into some new version of ourselves over the course of our lives.

From the perspective of neuroscience, identity is the brain's Ship of Theseus, and memory is the material that builds our identity. Say you were to somehow find my brain cells that were active when forming the memory of holding up my Pokémon card collection. And say you were to somehow keep track of which exact cells these were over the years. Every so often, you'd peer into my brain, find these particular cells again, and ask me to recall the memory. Some of these cells would turn on again for a period of time. I offer up this hypothetical because, in fact, scientists have observed something similar—just not quite as sci-fi. To start, the brain has neurons that respond selectively to pictures of people we know, including ourselves. These cells even respond if just the name of a person is written down or spoken out loud: if I see or hear "Pedro Ramirez" or look at a picture of my dad, the same cells will light up in activity for each. If I see or hear "Steve Ramirez," a different group of cells will become active. In other words, the brain contains neurons that represent the identities of individuals in our lives. At any given moment, every plank on the Ship of Theseus serves some function to keep the vessel moving forward.

The experiment is only halfway done though. If you kept looking at my brain cells that represent my Pokémon card collection over the years, you'd end up seeing something that has baffled scientists since they first were able to track such cells: at some point, and despite still being able to recall my Pokémon memory, an entirely *new* team of cells would be active. My engram of holding up my beloved Pokémon card collection has shape-shifted at the cellular level. Like the old wooden planks in the Ship of Theseus, they've been replaced with a completely new set of cells. Ongoing experiences reorganize old memories in the brain.

The phenomenon is called *representational drift*: the cellular representation of an experience changes, or *drifts*, over time. Neuroscientists have found that the brain represents even simple sensory stimuli in the world through drifting populations of cells. For instance, my brain processed the woody aromatics of my dad's cologne on his birthday with one group of neurons. That memory is almost certainly subserved by a completely different group of neurons today. Even if I were to actually smell my dad's cologne over and over again, the group of cells responsive to the cologne would change over time as some cells stop firing while others start firing to represent the experience. This change is so drastic that my cellular representation of this smell today likely looks nothing like it did on my dad's sixtieth birthday. The memory is stable but the very cells that produce it are not.

As experience accrues in the brain—as we live our lives—the neurons that processed a memory when it was formed begin to drop out while new ones come into play and take over. At a fundamental level, this means parts of my identity, as forged by my memories, drift more than others. I am both stable and unstable. It's this kind of cellular dynamism that sparks change in the psychological manifestation of the self. This inherent changeability is adaptive: the more we remember a particular memory, the faster its cellular representation can drift into something new and incorporate more information. Like any good sports team, drift is a way for the brain's roster of neurons to have healthy, active players on the court of experience who are ready to run a play, to adapt to unpredictable circumstances, or to be substituted in and out if needed, all with the common goal of surviving into the next round. It's a way for our brain's game plan to continuously adapt without being overly predictable or stuck in its ways.

The brain's physical machinery that enables memories is being continuously modified by time and experience. From synapses, to the wiring between cells, to the activity of teams of cells, our memories constantly renew these molecular and cellular pieces, like waves crashing on a beach, tidal in their composition and drifting in their very nature.

As we'll see next, the science of how stable memories emerge out of unstable biology is where we can learn about how the brain produces the different transformations we endure in life. It is the brain's dynamism that lets such transformations happen, for the only constant in the neuroscience of memory is that of change. My dad's sixtieth birthday signaled a new era in his life, and my Pokémon card collection symbolized the wide-eyed innocence surrounding my youth. After Xu's death, my identity, who I was as a person, had to transform once more, and the process of that transformation lead me into a spiral of addiction and all the way to a second chance at life. That night at Capital Grille marked the beginning of my next great transformation. And I'm deeply thankful for the brain's ever-changing physical configuration because it's the reason that I am here today. The brain's mutability enabled my well-being.

———

We've now explored how we imagine and dream about what our lives might be like in the future. And we've examined the ways in which memory makes these processes possible. Now we can look at how we actually live those lives—how we change and grow, especially in the face of adversity.

At a fundamentally biological level, the triumphs and personal struggles that underlie the story of science and friendship that I tell in this book live in my memory. There are at least three threads of memories throughout the book that tie together the person I am today—two at the forefront and one lingering in the background. The first is my memory of losing Xu, which still teaches me about the depths of friend-ship and loss. The second is my family, who still teach me about finding love amid such loss. The third—often only hinted at to reflect how ghost-like it can be in its visibility—is my memory of how alcohol com-mandeered my life for the better part of a decade. My experience with alcohol taught me that quiet grief can become a stick of dynamite for our physical and mental health when left unattended.

In the aftermath of Xu's death, my memories of Xu gave my waking self a precious hope, but our lack of future together robbed my evenings of rest. My brain's way of making sense of a world without Xu became filled with more mysteries than I could've ever predicted. Lucid dream-ing brought a perpetual state of fatigue: while I could bring Xu's engram back to life through a vivid dreamscape, on nights when these encoun-ters were just too emotionally draining to anticipate, I discovered that drinking in excess was the one surefire method I could use to prevent lucidity altogether. It anesthetized my lucid self to a dreamless oblivion where solace existed—a literal blackout of the mind. Without alcohol, I felt like I was forced to make a nightly trip to what Hamlet called "The undiscover'd country, from whose bourn / No traveler returns." But I would return every single morning, worn down by experiencing the im-possible. If we indeed undergo a number of great transformations over a lifetime, then this version of me badly needed one. But I was stuck.

Even though we often use it as a way to wind down or to get rid of some jitters, alcohol is known to get in the way of a good night's rest. REM sleep (the part of sleep when we're most likely to dream) becomes impaired, and the amount of sleep we get overall in a given night de-creases substantially. Alcohol also decreases how much of a given dream we can recall, which perhaps explains why I found its amnestic proper-ties so appealing. For years, my recurring thoughts of Xu's death became a trapdoor that led to an emotional rock bottom. Discovery is what I did as a scientist; recovery is what I needed to do as a person. The latter is only possible with change. Regardless of how slowly or quickly my Ship was changing, I needed to completely overhaul it from the ground up.

Death, anxiety, and addiction—you'd think this trifecta would make a person run immediately to the closest psychiatrist and rehabilitation center. But they have a way of folding the will into solitude before heal-ing can begin. These things are pretty damn insidious. I wasn't ready to let out my internal cries for help—if they went unheard, then they'd only be real to me, safe from judgment. Facing this fear of judgment meant traversing through my most painful memories so that they could transform into my greatest sources of mending a mourning mind.

I don't really know when my relationship with alcohol went from occasional to seemingly unstoppable. Maybe it was preordained in some genetic capacity: differences in brain structures have been found to be part of a larger set of preexisting risk factors for developing a substance use disorder, as one meta-study from the National Institutes of Health (NIH) recently reported. I was always the kind of person to stay for another round, to work an extra hour, to push myself past some glass ceiling. I felt like greatness was expected of me because it's what it'd take to catch up to a world with a head start. For me, being a first-generation scientist automatically came with the expectation of doing something extraordinary. Stopping wasn't an option I ever really considered. Going full speed was my way of honoring my parents' sacrifices.

But it wasn't my career that my parents wanted me to nourish. It was my own life.

After Xu passed, I was a version of Theseus's Ship that was just trying not to capsize. When I was serene and not on a drinking binge, *when I was the calm*, I knew who I was, and I was proud to manifest Xu's dream through my research. The ship was sturdier. When I was dark and inebri-ated, *when I was the storm*, I felt alone, and I was ashamed of not having any real sense of control over my well-being. My identity was splinter-ing. But both extremes are nonetheless *me.* Every single one of my ex-periences consist of memories, good and bad, that are engraved throughout my identity, inseparable and inexorable in their influence. I am both the calm and the storm.

This admission is liberating—curative—because it lets the remem-bering part of my life aid the recovering part. The moment-to-moment fluctuations of what experiencing reality feels like has taken me from the valleys of death, anxiety, and addiction to the crests of well-being, self-care, and connection, where a peace exists. The panoramic view from this version of my Ship is restorative.

Remembering is a tool that can be used to break apart an identity or to build upon one. At this point, I had two major life events to make sense of—Xu's death and my drinking. They are both connected, down to the very synapses that enable my experiences of them. Alcohol changes the brain. It impairs activity in the circuits involved in emotion, memory, and stress. It produces long-term changes in virtually every kind of cell in the brain and, therefore, has direct access to molding the very organ that gives us an identity. So it makes sense that drinking would impact how I was processing Xu's death. Excessive rumination on emotionally daunting life experiences can spiral into bouts of depression and anxiety, for instance, and alcohol has the ability to temporarily alleviate negative feelings. When someone is in pain in the hospital, the doctor may offer a patient-controlled analgesia (PCA) pump, which, at the press of a button, will deliver pain relief. Wouldn't you press *off* on the worst feeling you've ever felt, even temporarily, if given the option?

By reminiscing about Xu and intoxicating myself habitually, I was restructuring who I was as a person. Life needed to make sense to me again—without Xu and now with an addiction to deal with, who was I becoming? I was trying to piece together what psychologists call our *narrative identity*, which is basically the story that we tell ourselves about who we really are. Constructing a narrative identity is a lifelong process, and ideally we want to be evolving toward some sense of self that feels unified and has purpose. Our narrative identity integrates our reconstructed past, perception of the present, and anticipated future into a cohesive account of our self. It is forged in the crucible of memory.

If our narrative identity is our own first-person story, threaded together by the events we experience, we are the main character and our exploration of the world is the plot. Like any story, we encounter twists in the form of heartaches and turning points that lead to triumphs. Somewhere hidden between all the things that can hurt us and everything we can learn, there is redemption. One of the goals of our narrative identity is to find this meaning, to find resolution amid adversity.

When discovered, we regain a sense of control and agency over our life story.

Embedded in the process of finding our narrative identity is the idea that painful life experiences can help us grow as people. This is a two-step process: first, we sit with the episode of life that hurt us. We remember it, we hit rewind on different parts to feel them again and again, and we explore how it affected us. Not surprisingly, these are the ingredients of therapy. In the second step, we take what we learned and use it for some form of good, either within ourselves or in the world, or both—the idea is that these good acts then become ingredients for happiness. Studies suggest that supporting other people can add meaning to our own lives because we find direct utility, or purpose, for enduring what we've endured. When we help other people, our actions give new meaning to the adversity we've faced. We can summarize the idea in this way: first focus on the memory, and then learn from it. How do we actually do this—how do we use memory to help change our perspective on a difficult life event?

Research on how people adapt to negative events shows that a sense of personal agency is one of the best predictors of psychological health. This makes intuitive sense: we want to feel like we are in control of creating our own narrative identity. Negative life experiences often remove this sense of agency; we then spend time and effort trying to rebuild it. Reframing memories of negative experiences into opportunities to reclaim our sense of agency can have profound effects on us. Change manifests in the brain at every level of analysis: from genes and synapses to cellular firing patterns and physical composition of circuits. All work together inside the brain to produce our sense of identity. As one subject in a 2017 clinical study, called "Searching for Happiness," wrote, "To really find happiness, you have to lose it first." This process of meaningfully resolving negative memories, in other words, can be a turning point in our lives.

Let me describe in more detail how this works, using a memory of mine that sticks out for me because of how much it has changed. It's gone from being an experience I used to just shrug off—*it is what it is*— to a true turning point in my life—*it is what I need it to be.* I used to share it with friends in a comedic and light-hearted tone:

"Oh my god, let me tell you about the time I fell into a bush like a dumbass! I was at a scientific conference in the Caribbean, and one night I got so messed up watching the Patriots in the Super Bowl. I met this dude from Massachusetts, who bought me ridiculously expensive whiskey, nonstop for good luck. And the Patriots won! On the way back to my hotel room, I tripped over a curb and ended up becoming best friends with a shrub. I actually managed to rally the next day with some mimosas. And, hey, at least I survived the bender."

Coming to terms with addiction has meant sitting with this memory, recalling it in full, while letting its painful details play out in my brain. Let's remember it now without skipping over those details.

In 2017, I was at a conference in Grenada. My room was at a beautiful beachside resort where the restaurants are on the sand and the margaritas are in full supply. When the day's conferencing ended, I headed to the bar with some colleagues to catch up and to watch the Patriots in the Super Bowl. Most of my colleagues left by halftime. Toward the end of the game, it was just me and a stranger, who also happened to be from Massachusetts. He offered to buy me round after round, and I was stoked to watch my favorite team win while I reveled in the free drinks.

At some point, I inevitably began thinking about Xu, and the drinks kept pouring. After seven hours at the bar, I got up to walk back to my room, which was only two blocks away, and realized I had absolutely no control over my body. Shortly into my hazy journey, I tripped over a sidewalk and fell into a bush, about a foot from a barbed wire fence. I was breathing heavily but laughing at myself, while looking up to the night sky. I thought I could see Venus and Mars, so I stayed there. I wondered what Xu would think if he saw me like this. A local came by, helped me up, and walked me to my room. My memory ends there.

I woke up the next day in blood-soaked sheets, with napkins sticking to my torn knees and elbows, a pounding headache, and the kind of hangover-induced anxiety that put me on autopilot—back to the bar for a remedy. It wasn't just a remedy for my hangover that morning—I was also deep into the longer-term physical symptoms associated with alcohol withdrawal. When I ordered my first drink around nine in the morning, I remember not being able to grasp the beer bottle—not because of my shakiness or sweating, which were certainly apparent—but because I couldn't unclench either of my hands. They were rigid and stuck in fists, as if my blood was curdling underneath, poisoned by the night before. And when I tried to mimic playing the piano to stretch my fingers, excruciating, razor-like stabs shot through my hands, forcing my fingers back into a knot. This was the first time my withdrawal was so severe that my body felt like it was shutting down, *because it was*. Sitting with my thoughts was something I had learned to avoid really well—an idle mind became addiction's workshop—so I asked for two double shots of vodka and a mimosa, which within minutes uncurled my hands and put an invisible cloak back over my anxiety.

Determined not to fall in public again, I went to my hotel room alone with six beers, a fifth of whiskey, and a liter of vodka. There's no need to hide drinks if you're alone, so I spent the afternoon on the balcony of my room, overlooking the shore, counting the waves as they came in and out, while the clouds came and went.

By dinnertime, I invited an old friend over for cocktails and asked if he wanted to stay and party together, as we were heading back to Boston the following morning, and we were on the same flight. He agreed, and my binging continued. By midnight we were out of drinks, and I was too drunk to do anything other than sleep. The last thing I remember was my friend saying that he should stay to make sure I'd be okay, despite my repeatedly telling him I wouldn't miss the morning alarm. A dreamless sleep followed.

I woke up to what felt like someone punching both my chest and back, while at the same time pouring sand over the back of my head. The inside of my mouth felt numb, though I could feel something dripping off my lips. I wanted to gag but couldn't let anything out, and for a split second I thought I was lucid dreaming about being in the vacuum of space.

I tried to swallow, and then I panicked, realizing I couldn't breathe at all. My entire midsection tightened just enough to unclog my throat, forcing me to exhale what little air was left in my lungs as I dry-heaved in silence before I could inhale again. According to my friend, he woke up to the sounds of flat hums and gargles and thought

I was having a nightmare. But when he turned the light on, he saw that I was quietly choking on my own vomit. He rolled me to my side and, at some point, through some miracle of biology, my brain and body decided to live on.

This is a painful memory for me, and a clear example of what alcohol can do to the brain in as little as twenty-four hours. As I recall it, one way in which I process the memory is to ask myself, *What would have been my last thought if I'd died that night?*

Since I can never really know the answer, I have decided to instead use the question as a form of guidance on how to live: *think about what you'll want to be remembering while on your deathbed, and do more of that in life.* I know that when I die, I want my last thought to be of my parents and my partner, all contained in one mental frame—a never-expressed engram crystallized in time. This answer gives me a playbook for how to take control of my life. I want to be in control of my final memory, and addiction almost took that choice away from me.

All I could think about in the subsequent months was how close I was to having everything cut to black that evening. A drunken midnight nap and, *just like that,* I'd drift off to a memory. And that'd be it. No teary goodbyes by some hospital bed. No retirement or retrospection of my career. No marriage or seeing my nieces and nephews graduate. No Ramirez Lab discoveries or chances to mentor future generations of Team X. No dinners with my parents or birthday parties with my best friends or gatherings with my colleagues. No lucid dream in which I meet up with Xu one last time. No fairy-tale ending—just cold, coagulated memories and bruised knees, found one warm morning in a hotel room with a few empty bottles laying around, all as the waves came in and out, and the overcast sky came and went.

Each time I looked at this memory in the months and years after Grenada, I felt a thunderclap that filled a void with regret. I was sad that the person I used to be had exhibited so many signs that he needed help, both before and especially after Xu's death, yet waited years to seek it out. I was sad that spending an evening having too many drinks, used as an emotional palliative, wasn't even close to being an isolated incident. And I was sad that technicolor memories, once full of youthful levity, had changed into ones that were monochromatic and imbued with guilt. Gone was the careless sprightliness of my twenties.

My childhood best friends and partner had an informal intervention that ended up being the helping hand I needed to stand up while at rock bottom. The opposite of addiction is connection, journalist Johann Hari says powerfully, and the connections with the people I loved gave me an armor to fight and resuscitate my life. It was at last time for me to do what I knew how to do best: put in the work. Changing into something new meant uprooting memories that were rotting me and learning from their most corroded parts. Therapy and group meetings helped me find these parts. Through both, I realized just how scared I was of losing the people I loved. My memories of Xu dying had become a persistent reminder of life's inevitable end. The recollection of a negative memory has the power to contaminate other memories, and I realized that the memory of Xu's death had imbued other memories—of Xu and of my time in lab—with a deep sadness. (I can only imagine that what happens in rodents may happen in humans as well: the cells that hold onto a negative memory can begin linking up with neighboring cells and forcing them to contain the negative emotions too. Fear itself spreads and grows like a symbiote from cell to cell, memory to memory. So, for me, what once felt safe now felt terrifying.)

Trying to drown out my memories with alcohol made things nastier because alcohol has the power to heighten the negative features of a memory into a debilitating, anxious state. What's more, chronically ruminating on negative experiences can even begin to deteriorate the brain from within and precipitate pathological hallmarks of depression, dementia, and various cellular abnormalities—all of which, of course, I want to avoid.

Accepting that I was afraid of the loss that comes with death was a turning point for me in my battle against addiction. I realized that I couldn't reverse the tides of life, but I could lean into its transience and become mindful of how pristine a present moment can be. Since our awareness of the present is at the fulcrum of the past and future, every moment is a once-in-a-lifetime experience, for better or for worse. This sentiment helped me to feel more in control than a martini ever could.

As my mom would tell me, when it comes to life, *"todos somos pasajeros."* We are all just passengers.

I also learned for the first time that generations of my family had quietly struggled with alcohol use disorder, and this realization gave me the slightest bit of reprieve, knowing that I was a leaf on a far larger tree. The old AA saying—"One drink is too many and a thousand not enough"—made much more sense to me now. The ensemble of cells that made *me* were resistant to change for reasons that extended back generations before them—and I was determined to completely reconstruct their identity. It was surprisingly empowering. At the same time, I learned how many people in my family had also overcome their struggles with alcohol. I believe that if generational demons in the form of predispositions to addiction can be passed down, then generational angels in the form of resilience can be inherited as well. There is real strength, a biological upside, to what we are born with and mold over a lifetime.

On a psychological level, I was changing my memory, giving myself more control over my identity. Today we know that we can enact such change not only psychologically but also neuroscientifically in the lab. We can modify the cellular landscape of memory to promote health and well-being. One study from Jan Born's lab found that experimenters could regulate the emotional responses of a negative memory by pharmacologically suppressing stress hormones at the time the memory is retrieved. For instance, if you recall the negative memory of, say, a car accident or a divorce, while receiving a drug that inhibits the stress hormone cortisol, you may report a less-intense emotional response to the memory. This research highlights a promising way to help us cope with life's more debilitating experiences.

We also know that a negative memory can be changed on a cellular level if a subject is given the chance to frame the event in a new way. The specific cells processing the negative experience begin to physically lose their connections and strength, and even become disconnected from producing a change in behavior. In one study out of Tufts University, researchers found that a negative experience was actively suppressed by recalling a corresponding safe experience, which might be why certain forms of exposure therapy work so well. So as you recall that car accident or divorce memory, you might also recall a safer experience, such as all the times we drive without having an accident or the times where we've found a new love in life. The cells that represent your memory of these new experiences physically replace the cells that recall the older, more aversive ones.

A complementary angle involves recalling positive memories, or "savoring the past." Our positive memories are strongly intertwined with our sense of self and are part of our narrative identity. Their utility to our everyday life cannot be overstated: positive memories activate the brain's reward circuitry and associate feelings of *good* with other ongoing experiences. They can be strategically used to get us out of a bad mood; to increase how social we are; to increase our motivation levels, pain tolerance, and patience; and to think flexibly and creatively about the past, present, and future. These kinds of memories have the amazing ability to modify our brain and behavior.

Finding the positive meaning of memories can help improve everything from how children remember having their tonsils removed to boosting cognitive function in adults with dementia. The phenomenon of going back to the past has appropriately been called *reminiscence therapy*. A particularly relevant study in 2021 found that after losing a loved one, individuals benefited from recalling memories if they did so in ways that had "positive self functions"—in other words, if we used these memories to maintain our identity, to problem solve, and to prepare for one's own death. The act of remembering, in this case, restructures our identity to persist into a future without our loved one. Memories, therefore, can be used to modify memories.

We've demonstrated how this works on a neuroscientific level in the lab. Artificially activating the cells that harbor positive memories can permanently suppress negative emotional responses. Within the human brain, positive memories recruit areas of the cortex that can inhibit fear-producing areas like the amygdala and increase reward-related chemicals (such as dopamine). This can also be done noninvasively by giving a subject a rewarding experience, ranging from food to music, during the recollection of a negative memory to produce a similar effect. In rodents, neuroscientists have shown that activating positive memories completely reorganized the cellular underpinnings of a negative memory, dampening the negative behavioral effects of recalling that memory. Artificially activating a positive memory itself becomes a kind of antidote to treat the brain.

Doctors can also do this in the clinic by administering drugs that range from anxiolytics and antidepressants, to MDMA, psilocybin, and ketamine. All of these drugs can dampen maladaptive stress responses, helping to change the emotional intensity of an experience and render a memory more bearable. Neuroscientist Christina Kim and her lab were recently even able to activate the cells harboring "psychedelic engrams"—cells that were active while an animal was administered psilocybin, for instance—to produce anxiolytic effects in mice without the need for the drug to be continuously administered! It is exactly this kind of research that reignites the revolution for treating the maladies of our biology.

In terms of non-drug-based approaches, it turns out that dancing, walking, jogging, cognitive behavioral therapy, yoga, and any form of exercise in general can be particularly effective at treating disorders of the brain, which provides a promising, low-cost, accessible, and healthy habit-building avenue for our mental health professionals to consider. Exercise in particular remodels memory circuits of the brain by promoting the creation of new brain cells. These new cells then replace older cells, speeding up representational drift—the changing of cellular pieces that make up memory. These lines of research are all encouraging, and they converge on the idea that a myriad of experiences adaptively reorganize the brain.

————

There are many theories of how the brain heals, but they all have one thing in common: they all require change. My struggle with addiction only made sense to me when I could change my understanding of the experience, viewing it through the lens of my recovery. This process of producing positive change from a major life struggle is what psychologists call a redemption sequence. Redemption sequences offer a way for people to lastingly cope with, adapt to, and find resolution in adversity.

They also offer a new model for how the brain heals, called posttraumatic growth.

Posttraumatic growth is a way to empirically measure the outcomes of our redemption sequences. These outcomes can include improvements in interpersonal relationships (for example, increased empathy and sense of belonging), seeking out new possibilities (say, new interests and ways of living), personal strength (like increased resilience and confidence), spiritual change (deeper beliefs and philosophies of life), and an appreciation of life (such as increased altruism and gratitude).

Said another way, identity is a lifelong narrative containing passages of redemption sequences, and the resulting growth is very real and very measurable. In my case, after I progressed from addiction to sobriety, I became closer to my friends and family; I've picked up new hobbies to sustain my health; my life is enriched with more moments of mindfulness; and I now help other people struggling with addiction by helping to run meetings with the recovery community. This story helps make my life meaningful; posttraumatic growth keeps the score.

When researchers provided quantitative ways to measure the outcomes of posttraumatic growth (see the 1996 paper titled "The Posttraumatic Growth Inventory: Measuring the Positive Legacy of Trauma"), one surprising finding was that people who experienced personally traumatic events reported *more* positive change compared to people who hadn't undergone similar events. For instance, mothers whose newborns were treated for severe medical problems report some benefits in the form of strengthened bonds with their family and an added layer of appreciation for their child and for life. The same holds true for people who have lost both parents within a short time frame. Similarly, a survivor of a natural catastrophe may experience growth in several areas, including a deeper understanding of their self and their perceived role in the world.

My relationship with alcohol ended on February 18, 2021, ten days after the anniversary of Xu's passing. I felt like a leaf that had just fallen off a branch, untethered from addiction and liberated to drift in whichever direction sobriety took me, free of expectation and circumstance.

It was a decade-long sequence, redemptive in its narrative arc and cellular at its core.

Sobriety has helped me feel at home in my own mind and to even welcome its labyrinthian design—one full of revelations at each corner. For instance, while the pandemic didn't give me a drinking problem, it did reveal I was living with a substance use disorder. And while alcohol didn't give me relief from Xu's death, it did reveal I was living with a tremendous amount of anxiety over my own purpose as a son and scientist. And while anxiety didn't make me a broken person, it did reveal I was living with the condition of being human. It's the one condition we all share—the rest is nuance.

Today, when I think about my memories, like my memory of Grenada, I accept that they are part of my journey—one that took me to some serious depths before I could attempt to climb out for a second shot at life. The odyssey heading downwards is where I did my exploring, and I stumbled; the return trip upwards is where I reflect and recover. There is joy to be found in recovery and remembering. Both are opportunities to reframe our conscious past. By remembering mine, I'm learning how to forgive myself too.

We usually learn things first and remember them after. Here it's the other way around: I'm remembering first so I can learn again. My memories of losing Xu have changed from being full of grief to being providential, guiding my thoughts and behavior in the world with a newfound sense of gratitude for the time I had with him and the time I have with those I love. All roads in these remembrances lead to shareable truths.

The first truth from my journey is that a scientist can look like me. The second truth is that a person in recovery can look like me too. I accept the truth of who I've been and who I am so that I can continue to be *better* as a person. In this realization, I hope that even one reader struggling with any of life's endless calamities finds a source of comfort to know that I'm in the fight with you. *Be bold, and mighty forces will come to your aid.*

As my memories change—as my identity sheds its cocoons of the past and metamorphoses into today's version of me—I feel a profound sense of tranquility. Like a rainy day, remembering can have a cleansing effect.[2] The key is that when I now feel the heavier emotions surrounding Xu's death and my purpose as a person, I'm no longer avoiding them. Rather than reach for a bottle when grief and anxiety start bubbling into my consciousness, my group meetings have taught me that remembering Xu is a chance to rejuvenate my experiences with him from within. Xu will always be waiting for me in the past. The neuroscientist in me looks forward to the day when memories—like the last time I saw him at Green Street—change, yet again, into something with new meaning. Maybe into wisdom. The change is not so that my grief can shrink—*it never does*—but so that my life can grow around it. Biology destines us to live an ever-changing life.

My favorite answer to the question posed by Theseus's Ship (Is a ship that has all its components replaced, one by one, over the years still the same ship?) comes from a friend, who shared his answer with me over dinner one night. When I'd first stopped drinking, I'd been so scared of not being or having "fun" anymore—it's one of the most common fears I hear among people attempting to leave alcohol behind. Would anyone like the new person that I'd become? That night at dinner with my friend, I was still worried about this but trying my best to hide it.

We were deep in conversation about how much life had changed in the few years since we'd last seen each other, when the waiter approached our table and asked us for our drink order.

Before I could respond, my friend asked if they had any nonalcoholic options on the menu.

"Thanks for doing that. And for . . . for not judging me for being sober," I said.

"Well, you're still the same Steve to me. You're just not drunk anymore."

Ah yes—the answer to the riddle of whether or not I'm the same person is so simple: the answer is *addition by subtraction* (he says sarcastically, learnedly). I'll take it.

———

Change is biologically inevitable. With our growing understanding of neuroscience, we know today that change can lead the brain down a pathological road, such as into addiction, or change can promote health and well-being, such as when grief turns into growth.

So what does it mean to know that my identity is always changing? It means that asking Theseus whether or not his latest ship is the same as the old one is like asking me which version of me is *really me*: the person who was addicted to alcohol or the person who is now sober?

I think the answer requires a slight rephrasing of the question: Given that change is unavoidable, what is it *for*? Change on its own does not have a goal. Organisms move forward or backward; thoughts occur inwardly or are expressed outwardly; molecules combine and recombine; our cells are recycled and replaced—all simply because that's the way nature works. But to what end?

The brain is always changing by default. However, when any kind of biological change happens with a goal in sight, then we call this *progress*. I have changed as a person because the biological pieces that make up my identity have perpetually shape-shifted over the decades. I have progressed as a person because the goal of these changes is to restore myself with health and, by extension, with happiness. My memories change as a product of their drifting cellular components; when I give them a goal, or some kind of ethically motivated outcome that I believe truly matters, then the point of having memories is to guide me in my actions so that I may live on and share them along the way. With or without a destination in mind, the whole point of a ship is to be a vehicle for exploration.

8

So Long Lives This, and This Gives Life to Memory

MUCH OF this book has dealt with our changing biology. We are constantly changing ourselves, whether we are having a cup of coffee, for instance, or going for a run—every single experience that we go through in life modifies the brain, from its DNA to its interconnected networks of cells. Memory is no exception: recalling a memory changes how that individual memory is stored in our brains, and a newly reconsolidated memory will never be the same as it was when it was first formed. But memory is also something bigger, guiding our decision-making as we live out each day: Are there leftovers in the fridge? What's the best route to take to work today? Does our partner prefer the left or right side of the bed? Memory guides us as we make little changes—it's time to go grocery shopping, there's a traffic jam on Route 1, never get in bed from the left side—and memory guides us as we turn these little changes into bigger, broader patterns in our behavior. In short, memory paves the way forward for our continued and changing existence.

Keeping in full view the power of memory to change our lives, any focus on memory manipulation must have an objective goal: to harness this capacity for change for good, in the service of increasing our collective well-being. Thanks in part to science fiction, the idea of artificially changing our own memories might elicit uneasy feelings of a dystopic future where relationships are erased, identities are replaced, and governmental powers implant thoughts in our heads to mind-control society. But such a dystopic future can be readily prevented by maintaining a scientifically literate electorate and society in which a basic understanding of how memory works then leads to responsible policymaking, from the courtroom to the clinic. And as we'll see in the pages that follow, I'll make the case that memory manipulation inspires a kind of future where it is used for the betterment of humanity. So let's bring everything together in this final chapter through the lens of change in the field of neuroscience and cook up a bright and also realistic future of memory manipulation.

———

As I prepared to graduate from MIT and to start my own lab as an independent investigator, I was grappling with change on many levels. Memory manipulation was officially underway as a viable way to change the brain, and neuroscientists were grappling with the ethical implications of having such a powerful technology at our disposal. From a scientific standpoint, I wanted to start a lab that would change memories to alleviate pathologies of the brain and that would share this knowledge freely with the world. From a personal level, I was changing—growing—amid losing my friend and connecting more deeply with my parents. My relationship with Xu was turning into a new relationship with my memory of Xu, and I was about to discover how terrifyingly beautiful this process could be. In short, the future that I'd imagined was now moving forward but in a completely different way than I'd envisioned.

To graduate with a PhD, I had to complete a thesis, which was a written summary of all the engram work I had done at MIT, and successfully pass a thesis defense, which was an oral presentation of this work in front a selected committee of professors as well as a general audience. My mom and dad, brother and sister, aunts and uncles, childhood friends and lab mates, Susumu and other professors—basically anyone who was interested—attended my defense on July 22, 2015. I walked into a packed auditorium adjacent to our building's atrium. Howard Eichenbaum sat at the front and told me that this was my trip to the Oscars— "You deserve to be here today." My family sat together in one corner of the auditorium, and my best friends were scattered throughout, each excitingly mouthing, "AHH," when I looked in their direction. As soon as I got up to the podium, I straightened my jacket, gave my tie a nervous tug, and read my title slide.

"Illuminating the mental memoriam. A thesis by Steve Ramirez."

In in the opening portions of my defense, I thanked Team X for the last five years. We were all about to embark on the next stages of our careers. I was applying to the Harvard Society of Fellows program and for an Early Independence Award from the NIH that would let me "Skip the Postdoc" and go straight to professorship after graduate school. Xu's passing made me embrace a new motto of ambition spoken in the movie *Interstellar*: This is no time for caution.

I paced back and forth across the front of the auditorium and gave a brief history of the engram field, as well as the key findings from my papers with Xu, each supplemented with a personal memory to showcase his compassion as he mentored me during this time. I talked about the time at a Society for Neuroscience meeting when Xu and I gave a presentation together to announce the results from Project X to the broader research community. I told Xu I felt a bit woozy from being nervous and hungover. After muscling myself through my half of the talk, I pretended to take a phone call, while he took over and I ran to the bathroom to throw up. When I got back, the talk was over, and he had two Alka Seltzers ready to go. Another time I had to sleep in the lab overnight because I was moving apartments and was without a home for twenty-four hours. Xu decided to stay with me past midnight—he ordered pizza, and we used the time to work on our first paper. I wanted everyone in the audience to hear about the good times Xu and I had together rather than to focus strictly on Xu's absence.

"It's in the kingdom of our memories of Xu that he becomes timeless," I said to the audience. "Every time we remember him, we mentally unglue ourselves from the physics of *was*, *is*, and *will be*. We join him in the phenomenal plane of permanence that is memory."

In the closing remarks of my defense, despite the promise of all the groundbreaking advances, I acknowledged that my story simply couldn't have a happy ending. If memory breathes life back into those we have lost, at some point it has to exhale too. Everyone saw a black and white picture of Xu and me as my final slide. We all knew where this was going, as the science part and lighthearted storytelling part was over. For the first time since Xu's death, I needed to recount the day of his passing out loud because then, and only then, would I feel like my story at MIT had some kind of closure. My thesis defense was the first academic milestone I needed to achieve without Xu there. I was beside myself as I kept thinking, *What the fuck am I doing here?* while somehow building up the courage to speak.

I told the audience about the morning of February 12, and about my dad's response when I asked him if he had a second: "Mijo, for you, I have a second and forever." I felt like I was swallowing wet cement as I shared the words, and then I couldn't speak again, just like that morning.

Struggling to continue, I looked into the audience for the people who were there to support me, my family and friends. A single memory erupted, and its warm incandescence guided me back to the moment, to a way to finish my speech. It was a memory from before I was even born, as I would inherit it from my father in the form of what, in my opinion, is the brain's secret to immortality: a memory that is passed down.

I remembered my dad telling me his story of what it was like sneaking across the border. Without realizing it until that moment in the auditorium, I was minutes away from completing *my parents'* story because their dream was to raise children who could attend school and receive the education they never could. My mom worked a labor-intensive, full-time job and raised us, while my dad worked three jobs at a time to save enough money for us to go to school. To get here, he had crossed the border, was deported on his first try, then crossed again, and got to Los Angeles, then somehow earned enough money to bring my mom, my sister, and my brother to the States, moved to New York, then New Jersey, and finally to Boston, where I happened.

"Those were the best days of my life, mijo," he would say, and I'd ask him how in the world those were his *best* days.

"We could only afford a studio apartment for your mom, brother, and sister, but every night we got to have dinner together and go to sleep in the same room by telling stories and horsing around or listening to a radio your aunt gave us. When I'd wake up in the middle of the night, I'd look over, and the most important people in my life were there with me, sleeping safely and peacefully in one room. We weren't surviving a civil war anymore in El Salvador; we were making it here together. It was all I've ever needed."

Born out of his optimism and my mother's love were a son and daughter who were now lawyers; it was my turn to realize their wishes of raising an exceptionally educated family who could bring goodness to the world. My father's memory of crossing the border with his ripped jeans and T-shirt, a liter of Sprite, a few dollars, and, most importantly, his own dream for his family that kept him moving forward was all I needed to keep moving ahead in mine.

I was exactly where I needed to be. All my pain and all my insecurity crumbled under clenched sweaty palms. At that moment, I stood tall in front of my family as Dr. Ramirez.

"*Todo hombre puede ser, si se lo propone, escultor de su propio cerebro,*" I remember saying under my breath, quoting Ramón y Cajal. "Every human can be, if they want to, a sculptor of their own brain."

My parent's vision for coming to the States was complete, and it was time to start my own new journey.

I said, "Xu's life was a star shooting briefly across the night sky for a second and lasting forever. That's how long it takes for an engram to begin in the brain and how long, I believe, one lasts too. That's what I've learned here. Thank you, Xu."

I left the auditorium with a few new fancy letters after my name, which "no one can ever take away," my mom and dad reminded me.

The night after my dissertation defense, I was fortunate enough to make a few more memories, and when I need my spirits lifted, I often rewind to them. My mom and dad took me to the local Bertucci's that we went to often when I was in high school, and I walked into a room—*SURPRISE!*—with most of my aunts and uncles (over a dozen in total) and most of my cousins (a lot) eagerly waiting to celebrate. The day after, my childhood best friends invited me to dinner at a restaurant around the corner from our Cambridge apartment, and when we got back to our apartment—*SURPRISE!*—they had arranged for all my graduate school friends to be there to celebrate. We all then went to a bar with a dueling piano show in downtown Boston so that we could enjoy the high life for an evening and dance into each other's memories.

"Now *that's* the dream," my mom and dad joked on the phone, "but be sure to still wear your nice jacket and not the same clothes you wear every day to the lab."

The science journalist Sharon Begley once said, "Somewhere, something incredible is waiting to be known." I believe that incredible somethings are waiting to be known within ourselves as well. At my thesis defense, I found out just how enlivening our memories of those we love can be—so much so that they can quite literally push us to do the hardest thing we've ever done. They place us exactly when and where we need to be and then give us just enough of a nudge to keep moving forward.

————

The end of my time at MIT prompted me to think back to its beginning, and to appreciate how far the memory field had come since then. Xu and I joined the search for the engram beginning with our memory reactivation paper in 2012. In the years that followed, the engram field grew into an international community full of discoveries, with successful attempts at finding and controlling the neural substrates of memory to understand both how memory works and how we can control memory to fix the brain from within itself. In reviewing the field in 2019, cognitive neuroscientists Elizabeth Phelps and Stefan Hofmann would write, "Science fiction notions of altering problematic memories are starting to become reality." Phelps and Hofmann knew and acknowledged that the science of memory editing is much more complicated than fiction, but they were right that we were witnessing the beginning of a new era in the study of memory.

This new era has ushered in four major themes in the science of memory manipulation, through systems neuroscience research done primarily on rodents and cognitive psychological research done on humans. It's important to keep in mind that if a researcher can edit memories in the mouse brain in a lab, we should expect that one day a doctor will be able to do this in a human brain, albeit using less-invasive methods and approaches more tailored to improving the patient's quality of life. The first and second themes regard the most effective time to edit a memory, which is unsurprisingly when it is at the forefront of our minds: memories can be edited while they're being stored, and they can be edited while they're being retrieved. A third theme is that memories exist throughout the entirety of the brain, not just in one brain area. And fourth, memories can be created from scratch. Together, these themes pave the way for memory manipulation to have therapeutic value. We know when we should try to edit a memory; we know that we can access a memory through different parts of the brain; and we know that a memory can be engraved into the brain by an outsider.

Let's take a closer look at each of these themes. The first theme—memories can be edited while they're being *stored*—refers to the time when the bits and pieces of information present in the outside world hit our brain wiring to give cerebral flesh to an experience. Editing can go in two directions: enhancement or suppression. One way to enhance memories during their formation is to intervene to strengthen the connections between brain cells and to further synchronize their patterned activity, all within seconds to hours after learning. We can do this by administering drugs or through more invasive strategies such as brain stimulation. Essentially, what we are doing is making the communication between the brain cells that hold a memory more efficient.

In 2012, Xu and I were able to watch the mouse brain as it made a memory, and we were able to keep track of the cells involved, which opens up the opportunity to tinker with them and therefore manipulate memory itself. In humans, an experimenter can place a human subject in an fMRI and give verbal instructions ("Try to remember the list of pictures you're about to see," or "You're about to navigate a maze we've created, and your goal is to remember the route taken"). In this way, the researchers can collect brain scans while a memory is being made and then proceed to relate the brain's activity to any aspect of memory. They can even use the patterns of brain activity to decode various properties of memories, such as its emotional hue, strength, age, and accuracy.

Once a memory is accessed as it is being formed, it can be modulated and even strengthened by stimulating the human brain itself, which has been successful in both young and older adults. Transcranial direct current stimulation (tDCS), for instance, is an approach in which mild electrical stimulation is delivered through the scalp. Deep brain stimulation (DBS) is a more invasive approach in which electrodes implanted in the brain deliver small amounts of electricity. Both approaches have been used to disrupt or enhance our memories of facts and events. One influential study led by neuroscientist Nanthia Suthana and neurosurgeon Itzhak Fried found that stimulating brain areas near the hippocampus while a subject learned the locations of various landmarks enhanced their spatial memory, meaning that they were able to navigate to the landmarks faster on subsequent tests than they were before the stimulation. While the efficacy of brain stimulation depends on stimulation parameters and areas of the brain affected, it nonetheless offers a direct route to modify the various streams of information comprising a given memory.

A more indirect, albeit just as impressively effective, route involves altering the brain with drugs. Researchers in the Netherlands have administered pharmacological stimulants, including amphetamines and steroid hormones, such as glucocorticoids, to improve memory. Even increasing our sugar levels right before or right after forming a memory can lead to memory enhancement. And, of course, everyone's favorite thing to do, i.e., physical exercise, has been repeatedly found to improve memory and even to promote the growth of the hippocampus and the creation of new brain cells. None of this is science fiction anymore!

If we can enhance a memory by stimulating the brain directly via tDCS and DBS, or indirectly via drugs, then it makes sense that we could use the same process to weaken a memory as it is being formed. For instance, researchers have disrupted the proper functioning of the hippocampus by *over*stimulating it with DBS, which in turn impairs memory. What's more, certain drugs that block the synthesis of molecules important for memory are enough to produce amnesia. And there can be too much of a good thing: extreme bouts of exercise, such as running a marathon, can impair recently acquired episodic memories but improve other kinds of memories that do not require conscious episodic detail. As with life, balance for the brain is key.

So engrams can be activated, enhanced, and suppressed by using brain stimulation techniques, drugs, and behavioral strategies that successfully toggle the very cellular processes enabling memory.[1]

Our second theme is that engrams can be artificially edited while they're being *retrieved* or *recalled*. Just as with storing memories, retrieving them makes them vulnerable to all sorts of modifications, including strengthening, disruption, and incorporation of new information—and these have therapeutic value (as we saw in chapters 3–5). Drugs and behavioral interventions, or invasive genetic strategies, can be used to make memories stronger or weaker, which provides a second chance to turn the volume up or down on a given experience. A key difference is that unlike with storage, which only happens once for a given memory, we can access a given memory as many times as we'd like through artificial stimulation or active recollection. This means that there is enormous therapeutic potential in editing a memory while someone is recalling it. In humans, an experimenter can noninvasively achieve this feat by repeatedly accessing the given memory through verbal instructions ("Can you walk me through your last vacation?") or by presentation of its associated cues ("Here's a picture of your high school") and use this behavioral approach to access the memory, which renders it malleable and primed to become updated with new information. Remembering makes experience unstable so that we can restabilize it as we see fit.

Our third theme is that engrams are located everywhere in the brain. On the one hand, this makes targeting the entirety of a memory challenging: whether it's remembering that a tone predicts a shock or remembering the vivid episodic details of our wedding day, memories are distributed throughout the entirety of the brain. They engage every type of cell, brain area, and circuit that we know of, making the search for an engram feel more like trying to find a needle in a haystack made of needles. On the other hand, while memories are indeed complex, we may not have to find an entire memory to activate or erase one. There are cells in numerous brain regions and circuits whose activity is enough to spark or inhibit recollection of a memory. This means that we can activate only these cells, and they can jump-start the cascade of brain events leading to remembering (as we saw in chapters 2 and 3). It's exciting to think that while the research that Xu and I conducted has focused on areas like the hippocampus or amygdala, scientists have taken the search for the engram brain-wide and have been able to manipulate learning and memory by targeting virtually *any* brain area at will. Memories, as complex as they are, are accessible via a number of different brain regions, which will make psychiatric and neurodegenerative disorders more readily treatable (as we saw in chapter 5) since researchers now have many different "entry points" for accessing a given experience in the brain.

Finally, our fourth theme is that engrams can be artificially created in the brain—for the time being, only in nonhuman subjects. This last theme does feel like it belongs in science fiction. But in reality, as long as scientists continue to come up with new ways of modulating the brain both invasively and noninvasively, it is conceivable that we could create a memory of an experience without the need for experience per se. All of memory leaves a mark on the brain, and experience is the sculptor. With tools like optogenetics, researchers now hold the chisel and can delicately carve exquisitely precise patterns of cellular activity that build memory from the ground up, including implanting false memory in rodents, as Xu and I had previously done (as we saw in chapter 4).

I imagine that as our memory manipulation strategies become more sophisticated, the contents of memory that we're able to create in the brain will follow suit as well, and even reach a point where researchers will be able to *restore*, in humans, entire memories once thought to be lost. Fittingly, a unifying thread among all four themes is that modern neuroscience can be stranger than science fiction: memory editing in humans, in my opinion, will only underscore how malleable our conscious experience and recollections of the world are—and just how tied these are to our sense of being. In other words, it is not just memory that is subject to manipulation: our entire conscious existence is subjective, dependent on our perceptions and remembrances of things past.

———

During my last few months at MIT, I was hell-bent on starting the Ramirez Lab. I had no doubt that I wanted to be a professor of neuroscience, and Susumu had given me scientific carte blanche to experiment with all the lab's resources. He encouraged me to use that time to collect as much data as possible so that I could take it with me and begin to develop my own line of research.

The intellectual breathing room during my last months gave me time to prepare for the greatest all-or-nothing risk in my career—and I say that because I didn't have a backup plan. I reveled in having no safety net. I felt like a scientist sprinting on a rapidly disintegrating wire while doing the Lambada over the active volcano that is academia. First, I needed funding. Then, I needed to find a university that would hire me.

My opening roll of the dice was worth over $1.5 million and would be a golden ticket to financially secure my lab. I spent months cooped up during the day at Trident Café on Boston's Newbury Street and during the evening at Eastern Standard's bar in Kenmore Square working on my application for a prestigious Early Independence Award ("DP5") from NIH. The grant lets you go straight from graduate school to running your own lab—and the application required a twelve-page research vision for the next five years, five letters of recommendation to convince NIH that I didn't stink at science, and approximately a bajillion pages of supporting documents. After months of wanting to throw my computer into the Charles River, I made it to the final stage—an interview.

I was invited to the Four Seasons Hotel in Washington, DC, to meet with a panel of leading scientists from around the world. My job was to give them a five-minute pitch, followed by a brief Q&A, and that was it. A million and a half bucks and the idea that was the Ramirez Lab came down to me talking for 300 seconds. That's $5,000 a second, or roughly $1,000 a word. *Scared money don't make money.*

My pitch consisted of only three PowerPoint slides that were 100 seconds each.

"We're in the era of memory manipulation," I announced. And the goals of the Ramirez Lab were twofold: to use memory as a therapeutic intervention in animals showing symptoms associated with psychiatric and neurodegenerative disorders, and to create a map of every cell in the brain involved in memory.

My work with Xu set the conceptual foundation for the kind of research my lab would carry out. After showing that we could control, activate, and even alter the contents of memories, we could now start to pick apart and rebuild experiences in the brain. And we could do so as part of a larger international neuroscience community that was already working to manipulate memory.

Researchers worldwide had shown that the same cells Xu and I studied in the brain also underwent lasting physical changes as memories were formed and that silencing or erasing these cells compromised memory. Together, this was as close to *causality* as we've ever been in neuroscience, I argued, because we found cells that could turn memories on and off like a switch. My vision for the lab was already within reach: we could use that switch to permanently bring memories back that were seemingly lost due to dementia, and to dampen memories that are infiltrating the brain and giving rise to anxiety- and PTSD-related symptoms. And on a personal level, having a lab would let me continue my search for the engram, a search that began with Xu during my first year of graduate school.

From a technical standpoint, my lab would do these experiments by combining the most powerful technologies available, including optogenetics, with various new tools to listen in on brain cells as animals formed, recalled, and warped memories. But there remained an inconvenient truth to be reconciled before our work could be applied to humans. When using optogenetics, neuroscientists turn a laser on to stimulate any kind of brain cell, and the stimulation then induces all sorts of changes in cognition and behavior—when the light is on, magic happens as thoughts and actions are optogenetically restored. However, when the light is turned off, that magic subsides as the temporary "light-dependent" improvements disappear. In humans, by analogy, we know that drugs and therapy can take weeks to months to induce lasting changes in cognition and behavior, and a single dose or session of either one usually doesn't produce obvious lasting effects. The brain needs time to rewire itself to a healthy baseline. This lag between treatment and lasting change in humans would make studying the effects of any brain intervention much more difficult.

To circumvent this limitation, I argued, my lab would try to discover new ways of inducing permanent, preventative, and restorative changes in the mouse brain that could serve as blueprints for future experiments aiming to do similar feats in humans. Whereas many neuroscientists were tackling these issues through pharmacological and cognitive-behavioral approaches, I wanted to do so by accessing the brain cells holding onto a memory directly. For instance, my lab would start off by *repeatedly* reactivating positive memories optogenetically in the mouse to measure if this could prevent the memory from deteriorating. Similarly, we would repeatedly reactivate negative memories to measure if this could decrease a fear response over a long period of time. And finally, I envisioned experiments in which we could permanently restore memories thought to be lost due to dementia by repeatedly reactivating and *strengthening* their connections in the brain. The idea was that the repetition of reactivation via optogenetics in mice would mimic the repeated reactivation of memory in humans via less invasive techniques. With this strategy in mind, my lab would break down the complex processes of learning and memory to individual cells, while artificially modulating each to measure how we could, through repetition, change an animal's cellular activity and behavior for a lifetime. If successful, our treatments focusing on modulating memory to reprogram the brain could be used in humans.

When my five minutes were up, a brief question-and-answer session began; the initial focus was more pragmatic than scientific:

"How do you imagine starting your lab?" asked one of the scientists.

I had led enough researchers in graduate school that I felt ready to coach a team, who would become the next generation's Team X, in a collaborative, collegial, and open manner. My goal was to dream big, have fun, and get shit done (as a former LinkedIn CEO once put it). This was all a long-winded way of saying what would've been captured by simply telling the committee, "I imagine starting my lab by being like Xu," if they had ever known him.

"What would you say is the most important question in engram research?" chimed in another scientist.

In my opinion, one of the most fundamental questions in engram research is: What does a living memory look like in its entirety in the brain? While most engram studies had focused on modulating memories by targeting specific areas of the brain like the hippocampus, I wanted to create an interactive, three-dimensional map of the entirety of a specific experience in the brain, while monitoring how this experience changes as a result of disease. Memories are imbued with an extraordinary amount of information, such as sensory experiences and emotions, which all live as dynamic, physical, three-dimensional webs of activity in our brains. Memory isn't the sole province of the hippocampus; it's what the entire brain does. Being able to map out the neural substrates of a complete memory could let us monitor the physical changes that happen before and after any kind of neural deterioration or particular disorder emerges. Since these changes occur throughout the brain, a map of how brain areas communicate with each other under healthy circumstances could also allow us to measure when and how miscommunication occurs to give rise to a pathology.

But this isn't enough. Creating a map of the *structure* of memory would be like looking at a piano for the first time. While we would be able to see where every key is located, we wouldn't know much about which sounds they produce or what kinds of music they're capable of making when the keys are played in sequences. There are two ways to hear what a piano is really capable of: play it or have someone else play it. First, one could begin by pressing each key and testing out which patterns produce songs and which produce discordance. In contemporary memory research, this is like optogenetically activating the brain's neuronal keys in a particular arrangement that produces the music of memory. Xu and I were lucky enough to hit a couple of these keys in the hippocampus, and now a goal of memory research would be to play these keys throughout the whole brain and note how this affects cognition and behavior. I believe that the current state of the field is that we're still *only* playing "Twinkle, Twinkle, Little Star": we can cause an animal to freeze or to prefer a particular location in an environment, but the complex and vibrant nature of our rich and personally experienced episodes are perhaps more akin to Beethoven's Ninth, recruiting all of the brain's resources to create the symphony that is an engram. A corollary objective, then, would be to create a *functional* map of memory by pressing ("stimulating") each neuronal key throughout the brain with exquisite timing and precision. If achieved, then I believe we would find causality between patterns of neural activity and memory, thereby providing us with the musical sheets that comprise the mind. It would be a true principle of neuroscience.

A second way to hear the music of the brain would be to eavesdrop on its activity while it performs. This would require a live view of the activity from brain cells as they store and recall memories. Modern approaches are indeed enabling neuroscientists to advance brain imaging tools at an exponential pace: using tools like fMRI and high-resolution cellular imaging, the field has gone from being able to listen in on a couple of cells, to dozens and hundreds, and now to thousands and even millions of cells simultaneously. In order to create a functional map of how *all* the cells involved in a memory are communicating with each other, this would involve, at the very least, combining the genetic tools that Xu and I used with cutting-edge microscopes that can image brain cells at work. This approach allows researchers to make a record of the brain as the crescendo of an engram grows into consciousness, and to selectively listen in to the activity of brain cells and measure which aspect of memory they're producing. Like a conductor who has to understand and give shape to every sound produced from the orchestra to guide the symphony, neuroscientists can listen and find patterns to every bit of neural activity produced by the brain to guide recollection. By combining genetic tools with large-scale cellular imaging technologies, neuroscientists can detect when a pattern of keys or beats or rhythms in the brain are playing out of tune and consequently giving rise to a malady, and then combine what we know about the *structure* and *function* of each cellular instrument to ideally retune and restore their activity back to a healthy melodic state.

"What about in humans?" the same scientist asked.

The path to breakthroughs in human neuroscience is paved by ideas and technologies. For instance, one idea is to apply in humans what optogenetics taught us in rodents but use techniques that we know already work in the human brain, such as DBS and similar techniques to stimulate brain cells. In a stunning set of experiments from neuroscientists Meaghan Creed, Vincent Jean Pascoli, and Christian Lüscher, the group used optogenetics to abolish pathological behaviors in rodents, and then developed and refined optogenetic-inspired DBS procedures that could be tested in humans too. Everything we do in rodents can inspire innovative therapeutic interventions by using already-existing tools in the human brain. It's all connected: we need basic science in order to fuel applied science, and I wanted my lab to be right at the middle.

The panelists nodded approvingly (I think). Once I was done answering their questions, one of them simply said, "Good," which I think meant . . . good. When my five minutes were up, I went back to my hotel room and stared at the ceiling for a few minutes before calling my parents and telling them I was still in one piece-ish.

If my first bet was a roll of the dice to secure funding through NIH, my second bet was finding a home university with lab space where I could conduct my proposed research. At the same time that I was working on my DP5, I had prepared an application for the Harvard Society of Fellows. The Society had a mythical ethos around it. My understanding was that it consisted of extravagant dinners where Fellows wore clothes that were the result of Louis Vuitton and Rolex having a baby, they wore shoes made out of currency that John Harvard himself signed, and for dessert they ate Pulitzer Prizes, all while guffawing at people who don't use the word *guffaw*. I applied, though, because the Fellows position secured me three years of *do whatever you want to jump-start your career* time, and the *whatever* I wanted was to start the Ramirez Lab.

The interview was at the "Yellow House," the headquarters for all things Society, and I wore the same suit I wore for my DP5 interview for two reasons: for good luck and because it was the only damn suit I owned. I was still a graduate student, lest you forget. I made the mistake of biking to the Yellow House while it was still warm out, and I showed up looking like I had just finished last place in the Tour de France. I asked to use their bathroom to see the natural disaster that was my sweaty button-up, and I quickly began dousing water on the striped blotches of shirt that didn't have sweat on them, so as not to look like a dying zebra straight out of a National Geographic documentary.

When it was my turn to interview, I wiped the beads of sweat off my forehead, went into the interview room, and sat in front of a dozen Harvard professors seated in an intimidating, inverted-U shape, so that everyone could gaze directly into my soul. Unlike at NIH, here I had a *whopping* twenty minutes.

"Congrats on making it out of the Tonegawa Lab!" one professor began. He was referring to the lab's reputation of keeping grad students for a long time—a few researchers warned me about this before joining. I had graduated in the relative record speed of 5 years.

"Not without some scars to show for it," I replied chuckling, only half kidding.

"So, toying around with memories, eh? Isn't this the dystopia we see in every other sci-fi movie waiting to happen?" asked one professor, in what I thought was the most important question of the evening, and which we homed in on for the entire interview.

I explained that the key is prevention before we even need intervention, but we need to plan for both. If we have self-driving cars that don't crash, for instance, then maybe we don't need seatbelts; but the latter are almost certainly worth keeping because errors, no matter how rare, may still pop up. Memory manipulation and ethical dilemmas surely go hand-in-hand, and the goal is to prevent the misuse of memory manipulation while also planning to intervene if it falls into the wrong hands. This is easy to say and a lot harder to execute: if people can use something as elemental and nourishing as water to hydrate and survive, while also using it to waterboard someone and drain life away, then I posit that literally *anything* can be used for good or bad. This means we need a legal and social infrastructure to prevent such exploitation from happening in the first place, and these dialogues must occur between neuroscientists, lawyers, investors, teachers, engineers, psychiatrists, the public—anyone who has something at stake. And since we're talking about memory, that would mean *everyone* has a seat at the table.

Neuroscience is inherently interdisciplinary. It should be woven into the court of law, for instance, so as to promote forward-thinking rulings, especially with regard to memory-driven eyewitness testimonies. I believe that neuroscience should be sewn into our societal fabric as well. This means educating ourselves early on about the malleable nature of our recollections and how this flexibility can be leveraged to treat people with psychiatric and neurodegenerative disorders. It means training and adequately paying our science communicators to disseminate information responsibly. It means investing in both private and public organizations that provide financial support for neuroscience research. It means electing governmental leaders who are scientifically literate and don't just rely on research when they need it at the hospital; and it means voting out those who denounce basic science and can't see that the repercussions of their decisions will be felt in their lifetime and beyond. If we accept that neuroscience is based on reality and that it can work *for us*, regardless of whether or not someone else believes in it, then we have made the kind of progress that enables a forward-thinking, socio-legal machine, capable of defining, preventing, treating, and humanizing maladies of memory and the brain.

Memory editing is an inevitability. It can have beneficial uses in our society by existing as a potential medication rather than a procedure to be abused. As such, one solution to partially avert a dystopian future is to offer memory manipulation *only* in a clinical setting for starters, as opposed to recreationally à la *Total Recall*, the same way that a well-trained psychiatrist would only prescribe an antidepressant to a patient living with depression instead of to the entire population of Boston.

This view embraces the idea that the treatment of editing memory, when done responsibly, can be used like a drug that's administered responsibly. The faculty interviewing me shifted forward in their seats, apparently intrigued by this analogy. I explained that we have to personalize mental health treatments, like precision medicine, in a manner that puts the human first. After all, we don't want a black market–driven memory editing crisis on our hands that sequesters our humanity, so we concurrently have to plan ahead and outline what clinics and rehabilitation centers would look like. Rehabilitation centers would run research-based programs that use memory editing to serve patients and to provide a replacement for nonregulated (and potentially abusive) memory editing treatments. There would be tight regulations on how the treatment is administered. The existence of these rehabilitation centers means there would be less need for excessive law enforcement to prevent or stop misuse of memory editing treatments. Any accessible media (e.g., social, television)—as well as family training and counseling and community-based organizations—would be used to educate the consumer about this treatment. In these educational settings, the language used would not make a patient feel vilified, ostracized, or pejoratively labeled for seeking treatment. If all this sounds familiar, it's because these are lessons learned the hard way from our current opioid epidemic. Learning from history, it turns out, means learning from memory.

Twenty minutes flew by and my interview was done. I had officially thrown my second pair of dice on the table.

About a month later, the Society offered me a position as an incoming Fellow, and this meant my time in the Tonegawa Lab was officially over. With a big-kid salary (finally!), some lab space, and permission to hire my own staff, I had a new academic home starting January 1, 2016.

I still needed the NIH to pull through if I wanted to start (and fund) the lab. While waiting for their response, I was traveling frequently to give talks on our engram research and to gradually establish my own identity independent of the Tonegawa Lab. One conference brought me to Amsterdam, where I led a panel on memory manipulation in a breathtaking cathedral; I was surrounded by stained glass and heavy air, pungent from humid pews. After being awake for the entire red-eye, I got to my hotel room and melted into the bed with barely enough energy to eat the complimentary *stroopwafels*. Damn, those *stroopwafels* were good. I was half unconscious when my phone buzzed—it was an email from the NIH notifying me that I had been awarded the DP5. And this time, unlike when I learned that I'd been accepted to MIT, I didn't drop my phone into a bowl of cereal like a dumbass.

The greatest dual gamble of my career had paid off. Jackpot.
The Ramirez Lab could now materialize.

————

A year later, while standing in an offensively long, morning line at a Starbucks, my phone began buzzing with a barrage of social media notifications, as if someone had punched a beehive and stuffed it in my back pocket. I looked at my phone to see if we had finally discovered life on another planet and read, "dude WTF CONGRATULATIONS!" from one of my best friends. What was happening?

I scrolled through my notifications and saw that someone had posted a link to an announcement from the President of the United States. For our work on memory manipulation, the White House awarded me the Presidential Early Career Awards for Scientists and Engineers, which is the highest honor our government can give to researchers in the beginning stages of their careers. My first thought was that Xu and I would have won this award together. Suspended in the moment, I imagined that Xu and I would have gone to Washington, DC, with our families, and they'd watch us walk up to the stage and accept the award; and then we'd all celebrate together over some fancy dinner. It was a spectacular surprise for me and for Xu's memory, and one that was deeply personal for yet another reason.

The White House announcement is going to make my mom and dad freak out, I kept thinking, giddy to tell everyone in my family that I was going to the nation's capital for a ceremony. I immediately began checking flight prices so that I could surprise my parents with a trip for the ceremony—for over three decades, they'd wanted to take a picture in front of the White House but never had the opportunity. I texted them a snapshot of the announcement and told them I'd had no idea I was being considered for the award.

"'This is going to be in the Salvadorian newspapers, you know!" my dad texted back.

I saved up enough money to pay for my parents' full trip and was delighted to spoil them with a suite at the Hay-Adams (which is

Hey-Fancy) and a dinner at the Four Seasons (which cost a mortgage). I asked the chef to customize two plates: a lobster pot pie with brandied lobster crème for my dad, and a buttery white Chilean sea bass for my mom—their favorites.

"It's a small step above the Burger King we'd go to after soccer, mijo," my dad joked.

My mom pulled out her phone and asked if we could take selfies to send to our family in El Salvador.

"You're going to scare them into thinking something is wrong because you're wearing a tie," she teased.

We FaceTimed with my sister and brother, to playfully rub our meals in their faces, and decided we were long overdue for a family vacation.

"Punta Cana," my sister insisted, to no one's disagreement.

Toward the end of the evening, while waiting for banana puff pastries, my mom asked if we could watch my second TEDx talk with Xu. She didn't know I had never watched it. Xu and I gave our talk just a few months before he passed away, and the minutes on stage felt too painful to revisit. This would be the first time I viewed that experience as it actually occurred rather than viewing it from memory. I hesitated, and then opened my internet browser on my phone.

"*Si. Él debería estar aquí también.*" Yes, he should be here too.

I was ready to dissociate during the talk so that I wouldn't be visibly upset at dinner—sadness was the last thing I wanted to feel around my parents. In the first minute of our talk, Xu and I did a playful back-and-forth on stage, and I found myself laughing, ignited to the bone by his presence. I was connecting with him, closing my eyes to better hear the sound of his voice and letting each of his words carry me across time, back to a charmed moment on stage together. *What is grief, if not love persevering?*[2] Xu hadn't gone anywhere at all. I just needed to let him back into my life in a different way. When I opened my eyes, the talk was over, but Xu still remained, joining us for dinner and meeting my parents through the memories I spoke of. Every memory extends the life of the people within them.

When the dessert arrived, my mom reminded me that the venue where Xu and I spoke held a special place in her heart and memory. It was the same room in Faneuil Hall in downtown Boston where she and my dad were naturalized as US citizens. Her remark launched me backward and forward in memory—somewhere between my life with Xu and the rest of my life with my parents and beyond, I could see the person I was becoming, molded by their love into a citizen of this world. The thought left me resting my head on my mom's shoulders out of happiness, uplifted and grateful to have caught the memory we were making together. I thanked her for being the reason I was alive and promised to make some kind of impact on the world for the sake of goodness. With our memories together as my greatest source of strength, it's a promise I knew I could keep.

On our way back to the hotel, we stopped at the White House and took pictures of ourselves standing in front of its bright, milky freestones.

"Do you think we'll ever see a Latino president?" my dad asked.

"Of course," I said, "and it better not take generations of election cycles."

"It actually does look smaller than in pictures," he pointed out.

While scrutinizing its windows, he began talking sternly. "You know, if I'd stayed in El Salvador, I'm certain that I wouldn't have stopped working my way to the top. I was either going to become president or at least have a job working with the president in La Casa Presidencial. . . . When you get your first pair of shoes for finishing elementary school, you just want to keep running in them. And I did. As hard as it was telling this plan to my parents, I ended up leaving our village and got a job as a postman. I met your mom by delivering her mail. I became a banker and stayed involved in politics as much as I could. But I couldn't go to sleep knowing that the bullets we heard at night were houses being broken into, and that on the following day, I would have to shield your brother's eyes from the dismembered bodies put on display. It was no way for your mom, brother, and sister to live. So I ran to this country because of *la guerra* and, you know what, I've never regretted leaving. Look at us, we *still* got to the White House."

The next day, I received my award, and my parents took enough pictures to crash all social media servers. No matter the occasion, I knew Xu would be on such stages with me for the rest of my life.

When the ceremony was over and we were headed to the airport, I turned into a walking commercial and told my dad that since he and mom travel so much to El Salvador, they should enroll in Global Entry, a program that allows prescreened travelers to avoid security lines when returning to the States.

"It's awesome," I said. "I literally just put my fingerprints on a scanner, and I'm done and out of the airport."

My dad laughed. "Mijo, you're telling me I had to sneak into the country twice, and you get to just *waltz* right back in without a security check!?"

"Well, I mean, when you put it that way. . . . Yes, exactly," I said.

"I'll look into it," he replied.

———

Memories define and redefine who we are. They coalesce into an overall sense of being. When we relive the best memories of our lives, we glow. As the philosopher Dan Dennett once said in a TED talk, "The secret of happiness is: find something more important than you are and dedicate your life to it." To me, that something is threefold: studying engrams, Xu's memory, and my family.

When Xu and I started our experiments on manipulating memories, we often talked about how we wanted to change the culture of science by embracing collaboration at all levels. Let's all be a bit less wrong as an engram community, one data point at a time.

In the first year of opening my lab, I saw firsthand how influential our vision had really been for our field. What used to be one poster presentation in one meeting was over a hundred at several. What used to be a single talk was its own symposium full of scientists, all eager to share the sneak peak of an engram that their labs had discovered. "The brain is big enough for everyone," as Tomás Ryan reminds me. I find it beautiful that our engram community has come together to demystify how the brain's mental time machine works. I just wish one more person could enjoy that with me.

I believe that engram research will continue to change the world for the better, and the secret sauce is in the people who make up the field. One evening at a Society for Neuroscience meeting in San Diego, Team X and I held an event for over five hundred engram researchers at all stages of their careers—from undergraduates to graduate students, to postdoctoral fellows, to professors, and everyone interested in memory in general. It was an international celebration of engrams.

For Team X and me, it was also a personal day of remembrance. Everyone was moving on to the next stages of their careers, but that night, as each member of our forever Team X arrived and we all sat by the firepits overlooking San Diego Bay, now with our new lab members too, we remembered Xu, we imagined where our lives were going, and we dreamt of the possibilities. Our former mentors began to show up, as well as colleagues, collaborators, and researchers we had never met before, people new to the engram field, and scientists who had studied memories for decades. Within the first hour, we were at capacity. I felt myself forming an indestructible engram, mercurial but liberated from the passage of time, one that I could later share with everyone, through a true miracle of neuroscience.

I began walking around to meet people, and nearly everyone at some point commented that Xu would be proud to be there. Team X would be proud to have him there too. At the same time, I was learning to accept, and above all to respect, that there is a physical finish line at the end of every single life. It is up to us to choose how we *live* on one side and how we *live on* as a memory on the other. My quest as a neuroscientist to alter the past taught me how to change a memory, and it taught me how we become memory, the creator of worlds. These discoveries of science and life will always help in transforming grief into mindful moments to appreciate the fleeting feeling of being, magical and terrifying as the moments may be.

The gathering was jubilant: some scientists were eager to talk about the next wave of engram experiments, others wanted to talk politics, and still others wanted to keep the conversation light and talk about their recent travels. Throughout the night, I'd periodically sneak away to find Team X and ask them how the evening was going, which was my excuse to catch a social breather. When we brought up Xu, we remembered the first time we announced our findings for Project X half a decade earlier. For a brief moment, he was back with us.

Those brief moments fill life with purpose. Xu still walks *within* me. He's a companion who provides structure to the decisions I make about tomorrow and who is free to roam around my dreamscapes with me to do what we did best together: explore the unknown. When we advance into the unknown, we discover; and when we discover, our influence on the world becomes memorable. It is through discovery that I breathe life back into Xu, for a memory converted back into a life becomes a legacy.

Team X started with Team Xu, and he's now an engram worth living for too.

As the buzz grew beyond the rooftop, I heard a distant, deep laugh join in on our conversation. I knew it was him. I looked around and waited to hear it again, but nothing. I bowed my head and smiled. *Get it together, Steve.* Each time I would shuffle from group to group, I'd dart my eyes and search the venue, hoping to see a glimpse of him some-where—a shadow turning a corner, a figure hunched by the pool, a slen-der side profile wearing red flannel. I looked down to text a picture of the event to my parents and told them I'd call in a second when I found a quiet spot to talk, and then walked hastily through the crowd with my phone up against my ear so that I wasn't suddenly looped into a conver-sation. The sun was setting on a ruby-lit rooftop, and I stopped at its entrance, finally able to look up beyond the rows of people, hoping to see a valorous visitor and keenly focusing on vanishing all but one sil-houette and his voice.

When Xu laughed again, I hung up before my parents could answer, frozen by how real it felt.

The evening air was blissful. A breeze picked up, carrying with it the vicissitudes of a memory, and I pinched myself.

NOTES

Introduction: A Memory of a Memory

1. As neuroscientist Denise Cai told us over coffee one morning, "Build an army of people you trust. You need people to fight for you and to support you. Never forget that we belong to each other."

Chapter 2. The Shape-Shifter

1. In addition, this "unintended" region of the hippocampus—the one where we could in fact successfully reactivate memories—was the same region that another lab was studying. Michael Häusser's lab at University College London was also working on reactivating a memory in the hippocampus in mice. Xu and I initially chose to focus on a different area of the hippocampus to avoid direct overlap in our work, but mouse OF5 showed us that the place where the magic of memory reactivation happens was indeed the unintended area. I emailed Häusser to ask if their experiments had continued to work and to open up the possibility of a collaboration. One week later, we got a two-sentence response: "Thanks for your interest. We are still working on this project and hope to publish it soon—but it's not out there yet." Xu and I shared an ironclad principle of team-oriented science and continued to openly communicate our work to the field.

Chapter 3. Do You Want a Spotless Mind?

1. This fearlessness was popularly captured in the Oscar-winning documentary *Free Solo*, in which climber Alex Honnold was shown to have an underachieving amygdala in response to fear-evoking stimuli, such as climbing the face of a mountain with no equipment!

2. At the time, it was common for unofficial runners, called *bandits*, to jump in the race and put in some miles in a lighthearted act of solidarity (some were dressed festively, such as Sponge-Bob SquarePants or Super Mario). Still, the morality around running as a bandit, as I did in 2013, has understandably evolved in the last decade. I fully acknowledge this is a misguided approach for participating in official races. Registered runners have every reason to disapprove of some random person hopping in and potentially getting in the way, as well as using up the course's resources. So to the running community: I'm sorry, and please have mercy on me.

3. Preventing the return of fear in humans now had a basis in neuroscience. Their experiment went like this: on day 1, all human subjects received a visual stimulus in the form of a yellow square, which was paired with a mild shock to the wrist, or a blue square, which was not paired

with a mild shock to the wrist. On day 2, the subjects received a presentation of the yellow square and of the blue square (this served as a "reminder" session that one square predicted a shock and the other did not). To measure fear, the authors recorded the subjects' skin conductance response. In short, when you're scared, your fingers sweat a little more and conduct more electricity, which can be measured with electrodes and graphed as small blips of activity. The yellow square meant shock, and shock meant more sweat; and because the yellow square was paired with a shock on day 1, the presentation of the yellow square on day 2 led to an increase in the skin conductance response. The blue square didn't lead to such a response since it was never paired with a shock and acted as a baseline measure. After the reminder session on day 2, all subjects were divided into three groups: the first group was presented with the same two squares again repeatedly and without a shock 10 minutes after the reminder session. A second group received the same repeated presentations of each square but 6 hours after the reminder session. A third another group did not receive a presentation of the squares at all and acted as a control group. The authors found something striking: memories are modifiable but within a specific window of time. After the reminder session, the group that was presented repeatedly with the squares after 10 minutes permanently suppressed the associated fear memory, compared to the group that had recalled it after 6 hours or the group that hadn't recalled it all. Somewhere in the time between 10 minutes and 6 hours, a recalled memory is unstable and changeable.

4. The emotional memory that the scientists created on day 1 was the pairing of a visual stimulus—in this case a yellow square—with a mild shock to the wrist. On day 2, subjects would recall the memory and then try to "unlearn" the association between the stimulus and the shock by viewing the visual stimulus repeatedly *without* the shock attached.

5. This saying likely "evolved over time. A truncated version from William Wright appeared in 1887. An instance with different phrasing appeared 1891. An exact match attributed to Dugald Bell appeared in 1895. Future researchers may discover earlier instances. Martin Rees and Carl Sagan employed the saying and helped to popularize it many years after it was in circulation." See Quote Investigator, September 17, 2019, https://quoteinvestigator.com/2019/09/17/absence/.

Chapter 4. More to Remembering Than Truth

1. Previously, neuroscientists John Guzowski, Paul Worley, and their group developed a technique that allowed researchers to "infer the activity history of individual neurons at two times." They found that two different experiences recruited two different groups of hippocampus cells, while two similar experiences recruited mostly the same group of cells. These pioneering techniques and conceptual advances were pivotal for engram research, which sought to visualize the exact cells that were involved in specific memories; see Guzowski et al. 1999.

2. An fMRI machine is a giant, spinning magnet that can detect what's happening in your brain. It's a magnificent, million-dollar tool that owes its origins largely to a graceful scientific idea from Japanese researcher Seiji Ogawa in 1990: neurons are alive, they need oxygen, blood provides oxygen, and the flow of blood through the brain as it gives off oxygen also gives off a detectable signal. Blood without much oxygen gives off one kind of signal; fully oxygenated blood gives off a different signal. That difference itself can be used to measure changes in blood flow in the brain over time, which gives us a slow-motion capture movie of the living, breathing brain over time. Areas that are active—say, as you form a memory—use more blood, as they need more

oxygen. The slightly quieter corners of the brain use less blood and therefore soak up less oxygen. As fMRI memory researcher Maureen Ritchey from Boston College tells me, "[An fMRI] can find blood moving through your brain because molecules in the blood get pushed around by the magnet. And once the molecules are pushed, you can find their aftereffects."

3. Rewarding experiences are in abundance for rodents and can be used to alleviate maladaptive behaviors. For instance, Vanessa Gutzeit and Zoe Donaldson at the University of Colorado, Boulder, optogenetically manipulated social memories via the prefrontal cortex to buffer fear and anxiety responses. "One novel aspect of our study is that it suggests that harnessing neurons activated by positive social interactions may be useful in broadly ameliorating fear and/or anxiety," they presciently describe in their 2020 paper. Imagine being able to artificially activate a memory that you find genuinely pleasant and that you want to relive in its entirety so as to assuage feelings of anxiety at any given moment. This applies to social memories, new and old memories, fear and reward memories, and all of the varieties of experience that leave a mark on the brain.

Chapter 5. The Antidote from Within

1. As stated eloquently in Bill's soliloquy (played by David Carradine) from *Kill Bill: Vol. 2*, directed by Quentin Tarantino (Miramax Films, 2004).

2. Part of the diagnostic criteria for generalized anxiety disorder is that a person must display at least three of the following six symptoms: restlessness, feeling keyed up or on edge; being easily fatigued; difficulty concentrating or mind going blank; irritability; muscle tension; sleep disturbance (difficulty falling or staying asleep, or restless, unsatisfying sleep).

3. The negative valence systems involve processing fear, anxiety, and loss. Humans and nonhuman organisms are threatened by all three. The positive valence systems involve processing reward, how we evaluate reward, and the habits we form as a result. Humans and nonhuman animals enjoy sugar, drugs, and revisiting their favorite locations that previously contained something tasty. The cognitive systems involve understanding how attention works, how perception sculpts our experience, and how memory is realized within the brain. Humans remember where something good or bad happened and approach or avoid that place; nonhuman animals exhibit similar behavioral responses. Finally, the arousal systems deal with circadian rhythms, their relationship to sleep and wake cycles, and how these processes occur within the brain. Humans need to sleep; nonhuman animals sleep too, and the lack of sleep is detrimental to the brain and the body. The way in which the RDoC program proposed to study these phenomena ranges from genes and molecules and cells to neural circuits and behaviors and human self-reports. It is supposed to be the view from *within* for a psychiatric disorder, a neuro-centric view of the mind, complemented by the behavioral changes accompanying a given disorder.

4. These patterns can lead to the identification of "biomarkers," which are miniature, biological holy grails and the goals of the RDoC project. They are measurable changes in the body (including the brain) that indicate the presence of a pathology. For instance, in cardiac research, we know that high levels of cholesterol don't always mean heart failure, but they can begin to correlate with cardiac problems. We fittingly have dietary plans that limit the amount of certain types of cholesterol to prevent clogged arteries. But when it comes to the brain, we're just beginning to understand what biomarkers may look like, and this requires a full-on interrogation of the brain, which combines bottom-up and top-down approaches.

5. In 2012, neuroscientists Melissa Warden, Karl Deisseroth, and their team were attempting to alleviate depression-related symptoms in mice by using optogenetics. The authors began by looking in the prefrontal cortex, which is one of the most well-studied areas of the brain that is heavily involved in psychiatric disorders, consciousness, memory, attention—all the things that define us as humans. When the team activated the prefrontal cortex, much to their surprise, nothing happened. But, optogenetics lets us study both a brain region, like the prefrontal cortex, as well as the connections between brain areas by turning these connections on or off. As a reminder, a brain cell consists of a cell body; the dendrites, which receive incoming information from other cells; and the axons, which transmit information from one brain cell to the one that's next in line. Instead of activating the prefrontal cortex, Warden and her team wanted to see if axons originating from the same brain area can transmit different information depending on where these projections targeted. They discovered that stimulating the axons that originate in the prefrontal cortex and terminate into one area of the brain actually reduced depression-related behaviors. But, when they stimulated the axons that originate in the prefrontal cortex and terminate on a different area of the brain, the animals actually showed an increase in depression-related behaviors. It wasn't the origin of information that mattered as much as where that information was going. Think of it this way: your outgoing calls all originate from your cell phone, but the recipient of that call dictates whether the conversation is pleasant ("Hey, buddy, ol' pal!") or excruciating ("Yes, Comcast, I will hold for century.").

6. While we don't know the mental expanse of what animals can subjectively experience, we do know that we can at least measure neural activity and its relationship to more objective measures like the capacity to approach or avoid rewarding stimuli, weight changes, sleep-wake cycles, and various subsets of the proposed RDoC for classifying mental disorders.

7. Inspired by Marcel Proust, *In Search of Lost Time*, vol. 1, *Swann's Way* (Penguin UK, 2003).

8. In 2021, Speer and Delgado discovered that finding the "positive meaning" in memories of negative events could even make the negative memory less negative the next time it was recalled. As the authors note, "We show that positively reinterpreting negative memories adaptively updates them, leading to the reemergence of positivity at future retrieval. Focusing on the positive aspects after negative recall leads to enhanced positive emotion and changes in memory content during recollection one week later, remaining even after two months." In other words, finding the positivity in a negative memory adaptively changed how the memory can make us feel!

Chapter 6. A Second and Forever

1. For the neuroscientifically inclined reader, there's an abundance of cells in the brain identified by what they respond to, including time cells, grid cells, border cells, speed cells, and head-direction cells. For a review, see M. B. Moser et al. 2015.

2. In one fascinating rodent study, neuroscientists Karim Benchenane and Gaetan de Lavilléon at the French National Centre for Scientific Research forced a memory to be formed while animals slept: they stimulated the brain cells representing a particular corner of an environment while also stimulating brain cells that produced rewarding sensations, and they successfully connected the two experiences. This would be like stimulating a neutral memory in your

sleep—say, of a random corner of your living room—and connecting it with feelings of intense pleasure. Thereafter, you'd spend as much time as you could in that corner due to the rewarding feelings it conjures up.

Chapter 7. You Are What You Remember

1. All molecules turn over at different rates and across different tissues—some never do (e.g., enamel isotopes).

2. Weirdly enough, when I stopped drinking, my dreams occasionally became more about drinking. Even when lucid dreaming, I would look at a bottle of alcohol with immense guilt and take a sip, as if I was predestined to drink again. It turns out that a person's recovery correlates with how much they may even drink in a dream, and I experienced this firsthand: the further along in my recovery I am, the less that drinking appears in my dreams. In retrospect, maybe my brain was giving me a chance every night to grow beyond drinking. It was as if I could afford to relapse in my dreams so that I could feel the remorse that would happen and transfer this feeling to my waking life as a protective mechanism. These dreams, therefore, were dissociating alcohol from pleasure. Thankfully, the science out there hasn't found correlations between dream content and a risk of relapse; see Kelly and Greene 2019.

Chapter 8. So Long Lives This, and This Gives Life to Memory

1. For readers interested in the molecular biology of memory, recent fascinating studies have even suggested that brain cells can communicate with each other through virus-like mechanisms in which certain RNA that are important for memory are shuttled from one neuron to another for later use; see Pastuzyn et al. 2018.

2. From *WandaVision*, episode 8, "Previously On," directed by Matt Shakman, written by Laura Donney, Jac Schaeffer, and Peter Cameron, originally aired February 26, 2021, on Disney+.

REFERENCES

Introduction: A Memory of a Memory

Alexandra Kredlow, M., Fenster, R. J., Laurent, E. S., Ressler, K. J., & Phelps, E. A. (2022). Prefrontal cortex, amygdala, and threat processing: Implications for PTSD. *Neuropsychopharmacology, 47*(1), 247–259.

Eichenbaum, H. (2003). How does the hippocampus contribute to memory? *Trends in Cognitive Sciences, 7*(9), 427–429.

Eichenbaum, H. (2004). Hippocampus: Cognitive processes and neural representations that underlie declarative memory. *Neuron, 44*(1), 109–120.

Gruene, T. M., Flick, K., Stefano, A., Shea, S. D., & Shansky, R. M. (2015). Sexually divergent expression of active and passive conditioned fear responses in rats. *eLife, 4*, e11352.

Kandel, E. R., Dudai, Y., & Mayford, M. R. (2014). The molecular and systems biology of memory. *Cell, 157*(1), 163–186.

Maren, S., & Quirk, G. J. (2004). Neuronal mechanisms of fear and extinction: Implications for clinical disorders. *Nature Reviews Neuroscience, 5*(10), 844–852.

Parsons, R. G., & Ressler, K. J. (2013). Implications of memory modulation for post-traumatic stress and fear disorders. *Nature Neuroscience, 16*(2), 146–153.

Semon, R. W. (1904). *Die Mneme als erhaltendes Prinzip im Wechsel des organischen Geschehens.* Leipzig: Wilhelm Engelmann.

Semon, R. W. (1921). *The Mneme.* London: G. Allen & Unwin.

Semon, R. W. (1923). *Mnemic Philosophy.* London: Allen & Unwin.

Chapter 1. The Hundred-Year Quest for the Engram

Bird, C. M., & Burgess, N. (2008). The hippocampus and memory: Insights from spatial processing. *Nature Reviews Neuroscience, 9*(3), 182–194.

Boyden, E. S., Zhang, F., Bamberg, E., Nagel, G., & Deisseroth, K. (2005). Millisecond-timescale, genetically targeted optical control of neural activity. *Nature Neuroscience, 8*(9), 1263–1268.

Burgess, N., Maguire, E. A., & O'Keefe, J. (2002). The human hippocampus and spatial and episodic memory. *Neuron, 35*(4), 625–641.

Burnyeat, M. (1990). *The Theaetetus of Plato.* Hackett.

Corkin, S. (2002). What's new with the amnesic patient HM? *Nature Reviews Neuroscience, 3*(2), 153–160.

Corkin, S. (2013). *Permanent Present Tense: The Unforgettable Life of the Amnesic Patient, HM.* Basic Books.

Deisseroth, K., & Hegemann, P. (2017). The form and function of channelrhodopsin. *Science, 357*(6356), eaan5544.

Dudai, Y. (2010). The engram revisited: On the elusive permanence of memory. In *The Memory Process: Neuroscientific and Humanistic Perspectives,* ed. S. Nalbantian, P. M. Matthews, & J. L. McClelland. MIT Press.

Eichenbaum, H. (2017). The role of the hippocampus in navigation is memory. *Journal of Neurophysiology, 117*(4), 1785–1796.

Eichenbaum, H., Otto, T., & Cohen, N. J. (1994). Two functional components of the hippocampal memory system. *Behavioral and Brain Sciences, 17*(3), 449–472.

Goode, T. D., Tanaka, K. Z., Sahay, A., & McHugh, T. J. (2020). An integrated index: Engrams, place cells, and hippocampal memory. *Neuron, 107*(5), 805–820.

Josselyn, S. A., Köhler, S., & Frankland, P. W. (2015). Finding the engram. *Nature Reviews Neuroscience, 16*(9), 521–534.

Josselyn, S. A., Köhler, S., & Frankland, P. W. (2017). Heroes of the engram. *Journal of Neuroscience, 37*(18), 4647–4657.

Josselyn, S. A., & Tonegawa, S. (2020). Memory engrams: Recalling the past and imagining the future. *Science, 367*(6473), eaaw4325.

Kandel, E. R., Dudai, Y., & Mayford, M. R. (2014). The molecular and systems biology of memory. *Cell, 157*(1), 163–186.

Knierim, J. J. (2015). The hippocampus. *Current Biology, 25*(23), R1116–R1121.

Lashley, K. S. (1950). In search of the engram. *Society of Experimental Biology Symposium, 4, Psychological Mechanisms in Animal Behavior.* Cambridge University Press.

López-Muñoz, F., Boya, J., & Alamo, C. (2006). Neuron theory, the cornerstone of neuroscience, on the centenary of the Nobel Prize award to Santiago Ramón y Cajal. *Brain Research Bulletin, 70*(4–6), 391–405.

Mayford, M., Bach, M. E., Huang, Y. Y., Wang, L., Hawkins, R. D., & Kandel, E. R. (1996). Control of memory formation through regulated expression of a CaMKII transgene. *Science, 274*(5293), 1678–1683.

Mayford, M., Barzilai, A., Keller, F., Schacher, S., & Kandel, E. R. (1992). Modulation of an NCAM-related adhesion molecule with long-term synaptic plasticity in *Aplysia. Science, 256*(5057), 638–644.

Mayford, M., Siegelbaum, S. A., & Kandel, E. R. (2012). Synapses and memory storage. *Cold Spring Harbor Perspectives in Biology, 4*(6), a005751.

Nagel, G., Szellas, T., Huhn, W., Kateriya, S., Adeishvili, N., Berthold, P., . . . & Bamberg, E. (2003). Channelrhodopsin-2, a directly light-gated cation-selective membrane channel. *Proceedings of the National Academy of Sciences, 100*(24), 13940–13945.

Owen, A. M., Coleman, M. R., Boly, M., Davis, M. H., Laureys, S., & Pickard, J. D. (2006). Detecting awareness in the vegetative state. *Science, 313*(5792), 1402–1402.

Owen, A. M., Schiff, N. D., & Laureys, S. (2009). A new era of coma and consciousness science. *Progress in Brain Research, 177,* 399–411.

Penfield, W. (1968). Engrams in the human brain: Mechanisms of memory. *Proceedings of the Royal Society of Medicine, 61*(8), 831–840.

Ramsey, L. A., Koya, E., & van den Oever, M. C. (2023). Neuronal ensembles and memory engrams: Cellular and molecular mechanisms. *Frontiers in Behavioral Neuroscience*, 17, 1157414.

Schmolck, H., Kensinger, E. A., Corkin, S., & Squire, L. R. (2002). Semantic knowledge in patient HM and other patients with bilateral medial and lateral temporal lobe lesions. *Hippocampus*, 12(4), 520–533.

Silva, A. J., Paylor, R., Wehner, J. M., & Tonegawa, S. (1992). Impaired spatial learning in α-calcium-calmodulin kinase II mutant mice. *Science*, 257(5067), 206–211.

Squire, L. R. (2009). The legacy of patient HM for neuroscience. *Neuron*, 61(1), 6–9.

Squire, L. R., & Cave, C. B. (1991). The hippocampus, memory, and space. *Hippocampus*, 1(3), 269–271.

Squire, L. R., & Wixted, J. T. (2011). The cognitive neuroscience of human memory since HM. *Annual Review of Neuroscience*, 34(1), 259–288.

Tang, Y. P., Shimizu, E., Dube, G. R., Rampon, C., Kerchner, G. A., Zhuo, M., . . . & Tsien, J. Z. (1999). Genetic enhancement of learning and memory in mice. *Nature*, 401(6748), 63–69.

Thompson, R. F., & Donegan, N. H. (1986). The search for the engram. In *Learning and Memory: A Biological View*, ed. J. L. Martinez Jr. & R. P. Kesner, 3–52. Academic Press.

Tsien, J. Z. (2007). The memory code. *Scientific American*, 297(1), 52–59.

Tsien, J. Z., Chen, D. F., Gerber, D., Tom, C., Mercer, E. H., Anderson, D. J., . . . & Tonegawa, S. (1996). Subregion- and cell type–restricted gene knockout in mouse brain. *Cell*, 87(7), 1317–1326.

Chapter 2. The Shape-Shifter

Aharoni, D., Khakh, B. S., Silva, A. J., & Golshani, P. (2019). All the light that we can see: A new era in miniaturized microscopy. *Nature Methods*, 16(1), 11–13.

Barretto, R. P., Messerschmidt, B., & Schnitzer, M. J. (2009). In vivo fluorescence imaging with high-resolution microlenses. *Nature Methods*, 6(7), 511–552.

Barry, D. N., & Maguire, E. A. (2019). Remote memory and the hippocampus: A constructive critique. *Trends in Cognitive Sciences*, 23(2), 128–142.

Bontempi, B., Laurent-Demir, C., Destrade, C., & Jaffard, R. (1999). Time-dependent reorganization of brain circuitry underlying long-term memory storage. *Nature*, 400(6745), 671–675.

Cai, D. J., Aharoni, D., Shuman, T., Shobe, J., Biane, J., Song, W., . . . & Silva, A. J. (2016). A shared neural ensemble links distinct contextual memories encoded close in time. *Nature*, 534(7605), 115–118.

Flusberg, B. A., Cocker, E. D., Piyawattanametha, W., Jung, J. C., Cheung, E. L., & Schnitzer, M. J. (2005). Fiber-optic fluorescence imaging. *Nature Methods*, 2(12), 941–950.

Frankland, P. W., & Bontempi, B. (2005). The organization of recent and remote memories. *Nature Reviews Neuroscience*, 6(2), 119–130.

Ghosh, K. K., Burns, L. D., Cocker, E. D., Nimmerjahn, A., Ziv, Y., Gamal, A. E., & Schnitzer, M. J. (2011). Miniaturized integration of a fluorescence microscope. *Nature Methods*, 8(10), 871–878.

Goshen, I., Brodsky, M., Prakash, R., Wallace, J., Gradinaru, V., Ramakrishnan, C., & Deisseroth, K. (2011). Dynamics of retrieval strategies for remote memories. *Cell*, 147(3), 678–689.

Howard, M. W., Jing, B., Addis, K. M., & Kahana, M. J. (2007). Semantic structure and episodic memory. In *Handbook of Latent Semantic Analysis*, ed. T. K. Landauer, D. S. McNamara, S. Dennis, & W. Kintsch, 133–154. Psychology Press.

Kesner, R. P., & Hunsaker, M. R. (2010). The temporal attributes of episodic memory. *Behavioural Brain Research*, 215(2), 299–309.

Kitamura, T., Ogawa, S. K., Roy, D. S., Okuyama, T., Morrissey, M. D., Smith, L. M., ... & Tonegawa, S. (2017). Engrams and circuits crucial for systems consolidation of a memory. *Science*, 356(6333), 73–78.

Lei, B., Kang, B., Hao, Y., Yang, H., Zhong, Z., Zhai, Z., & Zhong, Y. (2025). Reconstructing a new hippocampal engram for systems reconsolidation and remote memory updating. *Neuron*, 113(3), 471–485.

Liu, X., Ramirez, S., Pang, P. T., Puryear, C. B., Govindarajan, A., Deisseroth, K., & Tonegawa, S. (2012). Optogenetic stimulation of a hippocampal engram activates fear memory recall. *Nature*, 484(7394), 381–385.

Madruga, B. A., Dorian, C. C., Sehgal, M., Silva, A. J., Shtrahman, M., Aharoni, D., & Golshani, P. (2024). Open-source, high-performance miniature multiphoton microscopy systems for freely behaving animals. *bioRxiv*, 2024.03.29.586663.

Mednick, S. C., Cai, D. J., Shuman, T., Anagnostaras, S., & Wixted, J. T. (2011). An opportunistic theory of cellular and systems consolidation. *Trends in Neurosciences*, 34(10), 504–514.

Moscovitch, M., & Gilboa, A. (2022). Has the concept of systems consolidation outlived its usefulness? Identification and evaluation of premises underlying systems consolidation. *Faculty Reviews*, 11.

Moscovitch, M., Nadel, L., Winocur, G., Gilboa, A., & Rosenbaum, R. S. (2006). The cognitive neuroscience of remote episodic, semantic, and spatial memory. *Current Opinion in Neurobiology*, 16(2), 179–190.

Nadel, L., & Hardt, O. (2011). Update on memory systems and processes. *Neuropsychopharmacology*, 36(1), 251–273.

Nadel, L., Winocur, G., Ryan, L., & Moscovitch, M. (2007). Systems consolidation and hippocampus: Two views. *Debates in Neuroscience*, 1, 55–66.

Ranganath, C., & Ritchey, M. (2012). Two cortical systems for memory-guided behaviour. *Nature Reviews Neuroscience*, 13(10), 713–726.

Ritchey, M., Libby, L. A., & Ranganath, C. (2015). Cortico-hippocampal systems involved in memory and cognition: The PMAT framework. *Progress in Brain Research*, 219, 45–64.

Rolls, E. T. (2000). Memory systems in the brain. *Annual Review of Psychology*, 51(1), 599–630.

Rubin, D. C. (2022). A conceptual space for episodic and semantic memory. *Memory & Cognition*, 50(3), 464–477.

Skocek, O., Nöbauer, T., Weilguny, L., Martínez Traub, F., Xia, C. N., Molodtsov, M. I., ... & Vaziri, A. (2018). High-speed volumetric imaging of neuronal activity in freely moving rodents. *Nature Methods*, 15(6), 429–432.

Squire, L. R. (2004). Memory systems of the brain: A brief history and current perspective. *Neurobiology of Learning and Memory*, 82(3), 171–187.

Teng, E., & Squire, L. R. (1999). Memory for places learned long ago is intact after hippocampal damage. *Nature*, 400(6745), 675–679.

Tomé, D. F., Sadeh, S., & Clopath, C. (2022). Coordinated hippocampal-thalamic-cortical communication crucial for engram dynamics underneath systems consolidation. *Nature communications, 13*(1), 840.

Tonegawa, S., Morrissey, M. D., & Kitamura, T. (2018). The role of engram cells in the systems consolidation of memory. *Nature Reviews Neuroscience, 19*(8), 485–498.

Tulving, E. (1985). How many memory systems are there? *American Psychologist, 40*(4), 385–398.

Winocur, G., & Moscovitch, M. (2011). Memory transformation and systems consolidation. *Journal of the International Neuropsychological Society, 17*(5), 766–780.

Yonelinas, A. P., Ranganath, C., Ekstrom, A. D., & Wiltgen, B. J. (2019). A contextual binding theory of episodic memory: Systems consolidation reconsidered. *Nature Reviews Neuroscience, 20*(6), 364–375.

Ziv, Y., Burns, L. D., Cocker, E. D., Hamel, E. O., Ghosh, K. K., Kitch, L. J., . . . & Schnitzer, M. J. (2013). Long-term dynamics of CA1 hippocampal place codes. *Nature Neuroscience, 16*(3), 264–270.

Chapter 3. Do You Want a Spotless Mind?

Alberini, C. M., & LeDoux, J. E. (2013). Memory reconsolidation. *Current Biology, 23*(17), R746–R750.

Alfei, J. M., Miller, R. R., Ryan, T. J., & Urcelay, G. P. (2025). Rethinking memory impairments: Retrieval failure. *Psychological Review,* https://doi.org/10.1037/rev0000538

Bird, C. M., & Burgess, N. (2008). The hippocampus and memory: Insights from spatial processing. *Nature Reviews Neuroscience, 9*(3), 182–194.

Bocchio, M., Nabavi, S., & Capogna, M. (2017). Synaptic plasticity, engrams, and network oscillations in amygdala circuits for storage and retrieval of emotional memories. *Neuron, 94*(4), 731–743.

Bolsius, Y. G., Heckman, P. R., Paraciani, C., Wilhelm, S., Raven, F., Meijer, E. L., . . . & Havekes, R. (2023). Recovering object-location memories after sleep deprivation-induced amnesia. *Current Biology, 33*(2), 298–308.

Cruz, F. C., Koya, E., Guez-Barber, D. H., Bossert, J. M., Lupica, C. R., Shaham, Y., & Hope, B. T. (2013). New technologies for examining the role of neuronal ensembles in drug addiction and fear. *Nature Reviews Neuroscience, 14*(11), 743–754.

Debiec, J., LeDoux, J. E., & Nader, K. (2002). Cellular and systems reconsolidation in the hippocampus. *Neuron, 36*(3), 527–538.

Denny, C. A., Kheirbek, M. A., Alba, E. L., Tanaka, K. F., Brachman, R. A., Laughman, K. B., . . . & Hen, R. (2014). Hippocampal memory traces are differentially modulated by experience, time, and adult neurogenesis. *Neuron, 83*(1), 189–201.

Drew, M. R., & Brockway, E. T. (2019). Regulation of fear extinction and relapse by hippocampal engrams. *Neuropsychopharmacology, 45*(1), 228.

Dunsmoor, J. E., Niv, Y., Daw, N., & Phelps, E. A. (2015). Rethinking extinction. *Neuron, 88*(1), 47–63.

Ergorul, C., & Eichenbaum, H. (2004). The hippocampus and memory for "what," "where," and "when." *Learning & Memory, 11*(4), 397–405.

Guskjolen, A., Kenney, J. W., de la Parra, J., Yeung, B. R. A., Josselyn, S. A., & Frankland, P. W. (2018). Recovery of "lost" infant memories in mice. *Current Biology, 28*(14), 2283–2290.

Han, J. H., Kushner, S. A., Yiu, A. P., Hsiang, H. L., Buch, T., Waisman, A., ... & Josselyn, S. A. (2009). Selective erasure of a fear memory. *Science, 323*(5920), 1492–1496.

Hsiang, H.L.L., Epp, J. R., van den Oever, M. C., Yan, C., Rashid, A. J., Insel, N., ... & Josselyn, S. A. (2014). Manipulating a "cocaine engram" in mice. *Journal of Neuroscience, 34*(42), 14115–14127.

Kensinger, E. A., & Corkin, S. (2004). Two routes to emotional memory: Distinct neural processes for valence and arousal. *Proceedings of the National Academy of Sciences, 101*(9), 3310–3315.

Khalaf, O., Resch, S., Dixsaut, L., Gorden, V., Glauser, L., & Gräff, J. (2018). Reactivation of recall-induced neurons contributes to remote fear memory attenuation. *Science, 360*(6394), 1239–1242.

Koya, E., Golden, S. A., Harvey, B. K., Guez-Barber, D. H., Berkow, A., Simmons, D. E., ... & Hope, B. T. (2009). Targeted disruption of cocaine-activated nucleus accumbens neurons prevents context-specific sensitization. *Nature Neuroscience, 12*(8), 1069–1073.

Kroes, M. C., Schiller, D., LeDoux, J. E., & Phelps, E. A. (2016). Translational approaches targeting reconsolidation. *Translational Neuropsychopharmacology,* 197–230.

LaBar, K. S., & Cabeza, R. (2006). Cognitive neuroscience of emotional memory. *Nature Reviews Neuroscience, 7*(1), 54–64.

Lacagnina, A. F., Brockway, E. T., Crovetti, C. R., Shue, F., McCarty, M. J., Sattler, K. P., ... & Drew, M. R. (2019). Distinct hippocampal engrams control extinction and relapse of fear memory. *Nature Neuroscience, 22*(5), 753–761.

LeDoux, J. (2003). The emotional brain, fear, and the amygdala. *Cellular and Molecular Neurobiology, 23*, 727–738.

LeDoux, J. E. (2020). Thoughtful feelings. *Current Biology, 30*(11), R619–R623.

LeDoux, J. E., & Lau, H. (2020). Seeing consciousness through the lens of memory. *Current Biology, 30*(18), R1018–R1022.

Lee, J. L., Nader, K., & Schiller, D. (2017). An update on memory reconsolidation updating. *Trends in Cognitive Sciences, 21*(7), 531–545.

Milad, M. R., Orr, S. P., Pitman, R. K., & Rauch, S. L. (2005). Context modulation of memory for fear extinction in humans. *Psychophysiology, 42*(4), 456–464.

Milad, M. R., Pitman, R. K., Ellis, C. B., Gold, A. L., Shin, L. M., Lasko, N. B., ... & Rauch, S. L. (2009). Neurobiological basis of failure to recall extinction memory in posttraumatic stress disorder. *Biological Psychiatry, 66*(12), 1075–1082.

Milad, M. R., & Quirk, G. J. (2002). Neurons in medial prefrontal cortex signal memory for fear extinction. *Nature, 420*(6911), 70–74.

Milad, M. R., & Quirk, G. J. (2012). Fear extinction as a model for translational neuroscience: Ten years of progress. *Annual Review of Psychology, 63*(1), 129–151.

Misanin, J. R., Miller, R. R., & Lewis, D. J. (1968). Retrograde amnesia produced by electroconvulsive shock after reactivation of a consolidated memory trace. *Science, 160*(3827), 554–555.

Nader, K. (2015). Reconsolidation and the dynamic nature of memory. *Cold Spring Harbor Perspectives in Biology, 7*(10), a021782.

Nader, K., Schafe, G. E., & Le Doux, J. E. (2000). Fear memories require protein synthesis in the amygdala for reconsolidation after retrieval. *Nature, 406*(6797), 722–726.

O'Leary, J. D., Bruckner, R., & Ryan, T. J. (2024). Natural forgetting reversibly modulates engram expression. *eLife, 13*.

Orsini, C. A., & Maren, S. (2012). Neural and cellular mechanisms of fear and extinction memory formation. *Neuroscience & Biobehavioral Reviews, 36*(7), 1773–1802.

O'Sullivan, F. M., & Ryan, T. J. (2024). If engrams are the answer, what is the question? In *Engrams: A Window into the Memory Trace*, ed. J. Gräff & S. Ramirez, 273–302. Springer Nature.

Perusini, J. N., Cajigas, S. A., Cohensedgh, O., Lim, S. C., Pavlova, I. P., Donaldson, Z. R., & Denny, C. A. (2017). Optogenetic stimulation of dentate gyrus engrams restores memory in Alzheimer's disease mice. *Hippocampus, 27*(10), 1110–1122.

Phelps, E. A., & Anderson, A. K. (1997). Emotional memory: What does the amygdala do? *Current Biology, 7*(5), R311–R314.

Phelps, E. A., & Hofmann, S. G. (2019). Memory editing from science fiction to clinical practice. *Nature, 572*(7767), 43–50.

Poll, S., Mittag, M., Musacchio, F., Justus, L. C., Giovannetti, E. A., Steffen, J., . . . & Fuhrmann, M. (2020). Memory trace interference impairs recall in a mouse model of Alzheimer's disease. *Nature Neuroscience, 23*(8), 952–958.

Poo, M. M., Pignatelli, M., Ryan, T. J., Tonegawa, S., Bonhoeffer, T., Martin, K. C., . . . & Stevens, C. (2016). What is memory? The present state of the engram. *BMC Biology, 14*, 1–18.

Quirk, G. J. (2002). Memory for extinction of conditioned fear is long-lasting and persists following spontaneous recovery. *Learning & Memory, 9*(6), 402–407.

Ryan, T. J., & Frankland, P. W. (2022). Forgetting as a form of adaptive engram cell plasticity. *Nature Reviews Neuroscience, 23*(3), 173–186.

Ryan, T. J., Ortega-de San Luis, C., Pezzoli, M., & Sen, S. (2021). Engram cell connectivity: An evolving substrate for information storage. *Current Opinion in Neurobiology, 67*, 215–225.

Ryan, T. J., Roy, D. S., Pignatelli, M., Arons, A., & Tonegawa, S. (2015). Engram cells retain memory under retrograde amnesia. *Science, 348*(6238), 1007–1013.

Salery, M., Godino, A., & Nestler, E. J. (2021). Drug-activated cells: From immediate early genes to neuronal ensembles in addiction. *Advances in Pharmacology, 90*, 173–216.

Sangha, S., Diehl, M. M., Bergstrom, H. C., & Drew, M. R. (2020). Know safety, no fear. *Neuroscience & Biobehavioral Reviews, 108*, 218–230.

Schiller, D., Monfils, M. H., Raio, C. M., Johnson, D. C., LeDoux, J. E., & Phelps, E. A. (2010). Preventing the return of fear in humans using reconsolidation update mechanisms. *Nature, 463*(7277), 49–53.

Schiller, D., & Phelps, E. A. (2011). Does reconsolidation occur in humans? *Frontiers in Behavioral Neuroscience, 5*, 24.

Sierra-Mercado, D., Padilla-Coreano, N., & Quirk, G. J. (2011). Dissociable roles of prelimbic and infralimbic cortices, ventral hippocampus, and basolateral amygdala in the expression and extinction of conditioned fear. *Neuropsychopharmacology, 36*(2), 529–538.

Silva, B. A., & Gräff, J. (2023). Face your fears: Attenuating remote fear memories by reconsolidation-updating. *Trends in Cognitive Sciences, 27*(4), 404–416.

Suzuki, A., Josselyn, S. A., Frankland, P. W., Masushige, S., Silva, A. J., & Kida, S. (2004). Memory reconsolidation and extinction have distinct temporal and biochemical signatures. *Journal of Neuroscience, 24*(20), 4787–4795.

Tonegawa, S., Pignatelli, M., Roy, D. S., & Ryan, T. J. (2015). Memory engram storage and retrieval. *Current Opinion in Neurobiology, 35*, 101–109.

Tulving, E. (1986). Episodic and semantic memory: Where should we go from here? *Behavioral and Brain Sciences, 9*(3), 573–577.

Tulving, E. (1987). Multiple memory systems and consciousness. *Human Neurobiology, 6*(2), 67–80.

Tulving, E. (1993). What is episodic memory? *Current Directions in Psychological Science, 2*(3), 67–70.

Whitaker, L. R., & Hope, B. T. (2018). Chasing the addicted engram: Identifying functional alterations in Fos-expressing neuronal ensembles that mediate drug-related learned behavior. *Learning & Memory, 25*(9), 455–460.

Zuniga, A., Han, J., Miller-Crews, I., Agee, L. A., Hofmann, H. A., & Drew, M. R. (2024). Extinction training suppresses activity of fear memory ensembles across the hippocampus and alters transcriptomes of fear-encoding cells. *Neuropsychopharmacology, 49*, 1872–1882.

Chapter 4. More to Remembering Than Truth

Abdou, K., Shehata, M., Choko, K., Nishizono, H., Matsuo, M., Muramatsu, S. I., & Inokuchi, K. (2018). Synapse-specific representation of the identity of overlapping memory engrams. *Science, 360*(6394), 1227–1231.

Cabeza, R., Rao, S. M., Wagner, A. D., Mayer, A. R., & Schacter, D. L. (2001). Can medial temporal lobe regions distinguish true from false? An event-related functional MRI study of veridical and illusory recognition memory. *Proceedings of the National Academy of Sciences, 98*(8), 4805–4810.

Coelho, C. A., Mocle, A. J., Jacob, A. D., Ramsaran, A. I., Rashid, A. J., Köhler, S., . . . & Frankland, P. W. (2024). Dentate gyrus ensembles gate context-dependent neural states and memory retrieval. *Science Advances, 10*(31), eadn9815.

Cowansage, K. K., Shuman, T., Dillingham, B. C., Chang, A., Golshani, P., & Mayford, M. (2014). Direct reactivation of a coherent neocortical memory of context. *Neuron, 84*(2), 432–441.

Gore, F., Schwartz, E. C., Brangers, B. C., Aladi, S., Stujenske, J. M., Likhtik, E., . . . & Axel, R. (2015). Neural representations of unconditioned stimuli in basolateral amygdala mediate innate and learned responses. *Cell, 162*(1), 134–145.

Guzowski, J. F., McNaughton, B. L., Barnes, C. A., & Worley, P. F. (1999). Environment-specific expression of the immediate-early gene *Arc* in hippocampal neuronal ensembles. *Nature Neuroscience, 2*(12), 1120–1124.

Ha, T-H. (2014). Think you've got a terrible memory? You don't know the half of it. (Interview with Elizabeth Loftus, Steve Ramirez, and Xu Liu). Ideas.TED.com, February 19, https://ideas.ted.com/think-youve-got-a-terrible-memory-you-dont-know-the-half-of-it/

Hirst, W., Phelps, E. A., Buckner, R. L., Budson, A. E., Cuc, A., Gabrieli, J. D., . . . & Johnson, M. K. (2009). Long-term memory for the terrorist attack of September 11: Flashbulb memories, event memories, and the factors that influence their retention. *Journal of Experimental Psychology: General, 138*(2), 161.

Jou, C., Hurtado, J. R., Carrillo-Segura, S., Park, E. H., & Fenton, A. A. (2023). On the results of causal optogenetic engram manipulations. *bioRxiv*, 2023.05.15.540888.

Kensinger, E. A., & Schacter, D. L. (1999). When true memories suppress false memories: Effects of aging. *Cognitive Neuropsychology, 16*(3–5), 399–415.

Kloft, L., Monds, L. A., Blokland, A., Ramaekers, J. G., & Otgaar, H. (2021). Hazy memories in the courtroom: A review of alcohol and other drug effects on false memory and suggestibility. *Neuroscience & Biobehavioral Reviews, 124*, 291–307.

Loftus, E. F. (1997). Creating false memories. *Scientific American, 277*(3), 70–75.

Loftus, E. F., & Klemfuss, J. Z. (2024). Misinformation—past, present, and future. *Psychology, Crime & Law*, 1–7.

Loftus, E. F., Miller, D. G., & Burns, H. J. (1978). Semantic integration of verbal information into a visual memory. *Journal of Experimental Psychology: Human Learning and Memory, 4*(1), 19.

Loftus, E. F., & Palmer, J. C. (1974). Reconstruction of automobile destruction: An example of the interaction between language and memory. *Journal of Verbal Learning and Verbal Behavior, 13*(5), 585–589.

Loftus, E. F., & Pickrell, J. E. (1995). The formation of false memories. *Psychiatric Annals, 25*(12), 720–725.

Ohkawa, N., Saitoh, Y., Suzuki, A., Tsujimura, S., Murayama, E., Kosugi, S., . . . & Inokuchi, K. (2015). Artificial association of pre-stored information to generate a qualitatively new memory. *Cell Reports, 11*(2), 261–269.

Patihis, L., Frenda, S. J., LePort, A.K.R., Petersen, N., Nichols, R. M., Stark, C.E.L., McGaugh, J. L., & Loftus, E. F. (2013). False memories in highly superior autobiographical memory individuals. *Proceedings of the National Academy of Sciences, 110*(52), 20947–20952.

Piantadosi, S. C., Zhou, Z. C., Pizzano, C., Pedersen, C. E., Nguyen, T. K., Thai, S., . . . & Bruchas, M. R. (2024). Holographic stimulation of opposing amygdala ensembles bidirectionally modulates valence-specific behavior via mutual inhibition. *Neuron, 112*(4), 593–610.

Pouget, C., Morier, F., Treiber, N., García, P. F., Mazza, N., Zhang, R., . . . & Vetere, G. (2024). Deconstruction of a memory engram reveals distinct ensembles recruited at learning. *bioRxiv*, 2024.12.11.627894.

Ramirez, S., Liu, X., Lin, P. A., Suh, J., Pignatelli, M., Redondo, R. L., . . . & Tonegawa, S. (2013). Creating a false memory in the hippocampus. *Science, 341*(6144), 387–391.

Redondo, R. L., Kim, J., Arons, A. L., Ramirez, S., Liu, X., & Tonegawa, S. (2014). Bidirectional switch of the valence associated with a hippocampal contextual memory engram. *Nature, 513*(7518), 426–430.

Reijmers, L. G., Perkins, B. L., Matsuo, N., & Mayford, M. (2007). Localization of a stable neural correlate of associative memory. *Science, 317*(5842), 1230–1233.

Schacter, D. L. (1999). The seven sins of memory: Insights from psychology and cognitive neuroscience. *American Psychologist, 54*(3), 182–203.

Schacter, D. L. (2001). *The Seven Sins of Memory: How the Mind Forgets and Remembers.* Houghton Mifflin.

Schacter, D. L. (2022). The seven sins of memory: An update. *Memory, 30*(1), 37–42.

Schacter, D. L., & Loftus, E. F. (2013). Memory and law: What can cognitive neuroscience contribute? *Nature Neuroscience, 16*(2), 119–123.

Schacter, D. L., Norman, K. A., & Koutstaal, W. (2000). The cognitive neuroscience of constructive memory. In *False-Memory Creation in Children and Adults*, ed. D. F. Bjorklund, 136–175. Taylor & Francis.

Slotnick, S. D., & Schacter, D. L. (2004). A sensory signature that distinguishes true from false memories. *Nature Neuroscience, 7*(6), 664–672.

Stott, R. T., Kritsky, O., & Tsai, L. H. (2021). Profiling DNA break sites and transcriptional changes in response to contextual fear learning. *PLOS ONE, 16*(7), e0249691.

Talbot, J., Convertino, G., De Marco, M., Venneri, A., & Mazzoni, G. (2024). Highly superior autobiographical memory (HSAM): A systematic review. *Neuropsychology Review*, https://doi.org/10.1007/s11065-024-09632-8.

Tonegawa, S., Liu, X., Ramirez, S., & Redondo, R. (2015). Memory engram cells have come of age. *Neuron, 87*(5), 918–931.

Vetere, G., Tran, L. M., Moberg, S., Steadman, P. E., Restivo, L., Morrison, F. G., ... & Frankland, P. W. (2019). Memory formation in the absence of experience. *Nature Neuroscience, 22*(6), 933–940.

Wing, E. A., Geib, B. R., Wang, W. C., Monge, Z., Davis, S. W., & Cabeza, R. (2020). Cortical overlap and cortical-hippocampal interactions predict subsequent true and false memory. *Journal of Neuroscience, 40*(9), 1920–1930.

Chapter 5. The Antidote from Within

Dunlop, B. W., Kelley, M. E., Aponte-Rivera, V., Mletzko-Crowe, T., Kinkead, B., Ritchie, J. C., ... & PReDICT Team. (2017). Effects of patient preferences on outcomes in the Predictors of Remission in Depression to Individual and Combined Treatments (PReDICT) study. *American Journal of Psychiatry, 174*(6), 546–556.

Fredrickson, B. L., Mancuso, R. A., Branigan, C., & Tugade, M. M. (2000). The undoing effect of positive emotions. *Motivation and Emotion, 24*(4), 237–258.

Haykin, H., Avishai, E., Krot, M., Ghiringhelli, M., Reshef, M., Abboud, Y., ... & Rolls, A. (2024). Reward system activation improves recovery from acute myocardial infarction. *Nature Cardiovascular Research, 3*(7), 841–856.

Hyman, S. E. (2012). Revolution stalled. *Science Translational Medicine, 4*(155), 155cm11–155cm11.

Insel, T. R. (2013). Transforming diagnosis. National Institute of Mental Health (NIMH), blog post, https://psychrights.org/2013/130429NIMHTransformingDiagnosis.htm.

Insel, T. R. (2014). The NIMH Research Domain Criteria (RDoC) project: Precision medicine for psychiatry. *American Journal of Psychiatry, 171*(4), 395–397.

Johnson, K. A., Okun, M. S., Scangos, K. W., Mayberg, H. S., & de Hemptinne, C. (2024). Deep brain stimulation for refractory major depressive disorder: A comprehensive review. *Molecular Psychiatry, 29*(4), 1075–1087.

Kim, S. Y., Adhikari, A., Lee, S. Y., Marshel, J. H., Mallory, C. S., Lo, M., ... & Deisseroth, K. (2013). Diverging neural pathways assemble a behavioural state from separable features in anxiety. *Nature, 496*(7444), 219–223.

McGrath, C. L., Kelley, M. E., Holtzheimer III, P. E., Dunlop, B. W., Craighead, W. E., Franco, A. R., . . . & Mayberg, H. S. (2013). Toward a neuroimaging treatment selection biomarker for major depressive disorder. *JAMA Psychiatry*, *70*(8).

Ramirez, S., Liu, X., MacDonald, C. J., Moffa, A., Zhou, J., Redondo, R. L., & Tonegawa, S. (2015). Activating positive memory engrams suppresses depression-like behaviour. *Nature*, *522*(7556), 335–339.

Scull, A. (2015). *Madness in Civilization*. Princeton University Press.

Sendi, M. S., Waters, A. C., Tiruvadi, V., Riva-Posse, P., Crowell, A., Isbaine, F., . . . & Mahmoudi, B. (2021). Intraoperative neural signals predict rapid antidepressant effects of deep brain stimulation. *Translational Psychiatry*, *11*(1), 551.

Siddiqi, S. H., Schaper, F. L., Horn, A., Hsu, J., Padmanabhan, J. L., Brodtmann, A., . . . & Fox, M. D. (2021). Brain stimulation and brain lesions converge on common causal circuits in neuropsychiatric disease. *Nature Human Behaviour*, *5*(12), 1707–1716.

Speer, M. E., Bhanji, J. P., & Delgado, M. R. (2014). Savoring the past: Positive memories evoke value representations in the striatum. *Neuron*, *84*(4), 847–856.

Speer, M. E., Ibrahim, S., Schiller, D., & Delgado, M. R. (2021). Finding positive meaning in memories of negative events adaptively updates memory. *Nature Communications*, *12*(1), article 6601.

Tye, K. M., Prakash, R., Kim, S. Y., Fenno, L. E., Grosenick, L., Zarabi, H., . . . & Deisseroth, K. (2011). Amygdala circuitry mediating reversible and bidirectional control of anxiety. *Nature*, *471*(7338), 358–362.

Warden, M. R., Selimbeyoglu, A., Mirzabekov, J. J., Lo, M., Thompson, K. R., Kim, S. Y., . . . & Deisseroth, K. (2012). A prefrontal cortex–brainstem neuronal projection that controls response to behavioural challenge. *Nature*, *492*(7429), 428–432.

Chapter 6. A Second and Forever

Addis, D. R., & Schacter, D. L. (2012). The hippocampus and imagining the future: Where do we stand? *Frontiers in Human Neuroscience*, *5*, 173.

Baird, B., Castelnovo, A., Gosseries, O., & Tononi, G. (2018). Frequent lucid dreaming associated with increased functional connectivity between frontopolar cortex and temporoparietal association areas. *Scientific Reports*, *8*(1), 17798.

Bendor, D., & Spiers, H. J. (2016). Does the hippocampus map out the future? *Trends in Cognitive Sciences*, *20*(3), 167–169.

Buckner, R. L. (2010). The role of the hippocampus in prediction and imagination. *Annual Review of Psychology*, *61*(1), 27–48.

Buckner, R. L., & Carroll, D. C. (2007). Self-projection and the brain. *Trends in Cognitive Sciences*, *11*(2), 49–57.

Carr, M. F., Jadhav, S. P., & Frank, L. M. (2011). Hippocampal replay in the awake state: A potential substrate for memory consolidation and retrieval. *Nature Neuroscience*, *14*(2), 147–153.

Comrie, A. E., Frank, L. M., & Kay, K. (2022). Imagination as a fundamental function of the hippocampus. *Philosophical Transactions of the Royal Society B*, *377*(1866), 20210336.

Damasio, A. R., Grabowski, T. J., Bechara, A., Ponto, L. L., Parvizi, J., & Hichwa, R. D. (2000). Subcortical and cortical brain activity during the feeling of self-generated emotions. *Nature Neuroscience, 3*(10), 1049–1056.

D'Argembeau, A., & Van der Linden, M. (2004). Phenomenal characteristics associated with projecting oneself back into the past and forward into the future: Influence of valence and temporal distance. *Consciousness and Cognition, 13*(4), 844–858.

De Lavilléon, G., Lacroix, M. M., Rondi-Reig, L., & Benchenane, K. (2015). Explicit memory creation during sleep demonstrates a causal role of place cells in navigation. *Nature Neuroscience, 18*(4), 493–495.

Dragoi, G. (2024). The generative grammar of the brain: A critique of internally generated representations. *Nature Reviews Neuroscience, 25*(1), 60–75.

Dragoi, G., & Tonegawa, S. (2011). Preplay of future place cell sequences by hippocampal cellular assemblies. *Nature, 469*(7330), 397–401.

Dresler, M., Wehrle, R., Spoormaker, V. I., Koch, S. P., Holsboer, F., Steiger, A., . . . & Czisch, M. (2012). Neural correlates of dream lucidity obtained from contrasting lucid versus non-lucid REM sleep: A combined EEG/fMRI case study. *Sleep, 35*(7), 1017–1020.

Duff, M. C., Kurczek, J., Rubin, R., Cohen, N. J., & Tranel, D. (2013). Hippocampal amnesia disrupts creative thinking. *Hippocampus, 23*(12), 1143–1149.

Eichenbaum, H. (2015). Does the hippocampus preplay memories? *Nature Neuroscience, 18*(12), 1701–1702.

Freud, S. (1899/1999). *The Interpretation of Dreams.* Trans. J. Crick. Oxford University Press. (Original work published in 1899.)

Gershman, S. J. (2017). Predicting the past, remembering the future. *Current Opinion in Behavioral Sciences, 17*, 7–13.

Gutzeit, V. A., Ahuna, K., Santos, T. L., Cunningham, A. M., Rooney, M. S., Zamora, A. M., . . . & Donaldson, Z. R. (2020). Optogenetic reactivation of prefrontal social neural ensembles mimics social buffering of fear. *Neuropsychopharmacology, 45*(6), 1068–1077.

Hartmann, E. (1998). Nightmare after trauma as paradigm for all dreams: A new approach to the nature and functions of dreaming. *Psychiatry, 61*(3), 223–238.

Hassabis, D., Kumaran, D., Vann, S. D., & Maguire, E. A. (2007). Patients with hippocampal amnesia cannot imagine new experiences. *Proceedings of the National Academy of Sciences, 104*(5), 1726–1731.

Hassabis, D., & Maguire, E. A. (2009). The construction system of the brain: Roles in navigation and memory. *Annual Review of Psychology, 60*, 1–24.

Konkoly, K. R., Appel, K., Chabani, E., Mangiaruga, A., Gott, J., Mallett, R., . . . & Caughran, B. (2021). Real-time dialogue between experimenters and dreamers during REM sleep. *Current Biology, 31*(7), 1417–1427.

Lane, R. D., Reiman, E. M., Ahern, G. L., Schwartz, G. E., & Davidson, R. J. (1997). Neuroanatomical correlates of happiness, sadness, and disgust. *American Journal of Psychiatry, 154*(7), 926–933.

Lisman, J., & Redish, A. D. (2009). Prediction, sequences and the hippocampus. *Philosophical Transactions of the Royal Society B: Biological Sciences, 364*(1521), 1193–1201.

Madore, K. P., Addis, D. R., & Schacter, D. L. (2015). Creativity and memory: Effects of an episodic-specificity induction on divergent thinking. *Psychological Science, 26*(9), 1461–1468.

Madore, K. P., Addis, D. R., & Schacter, D. L. (2017). Episodic specificity in future thought and memory. *Psychological Science, 28*(1), 1–10.

Madore, K. P., Szpunar, K. K., Addis, D. R., & Schacter, D. L. (2016). Episodic specificity induction impacts activity in a core brain network during construction of imagined future experiences. *Proceedings of the National Academy of Sciences, 113*(38), 10696–10701.

Maguire, E. A., & Mullally, S. L. (2013). The hippocampus: A manifesto for change. *Journal of Experimental Psychology: General, 142*(4), 1180.

Moser, E. I., Rowland, D. C., & Moser, M. B. (2015). Spatial representation in the hippocampal formation: A history. *Nature Neuroscience, 20*(11), 1448–1464.

Moser, M. B., Rowland, D. C., & Moser, E. I. (2015). Place cells, grid cells, and memory. *Cold Spring Harbor Perspectives in Biology, 7*(2), a021808.

Najib, A., Lorberbaum, J. P., Kose, S., Bohning, D. E., & George, M. S. (2004). Regional brain activity in women grieving a romantic relationship breakup. *American Journal of Psychiatry, 161*(12), 2245–2256.

Nguyen, N. D., Lutas A., Amsalem O., Fernando J., Ahn, A. Y., Hakim, R., . . . & Andermann, M. L. (2024). Cortical reactivations predict future sensory responses. *Nature, 625*(7993), 110–118.

O'Connor, M. F., & Seeley, S. H. (2022). Grieving as a form of learning: Insights from neuroscience applied to grief and loss. *Current Opinion in Psychology, 43*, 317–322.

O'Keefe, J. (1976). Place units in the hippocampus of the freely moving rat. *Experimental Neurology, 51*(1), 78–109.

O'Keefe, J., & Dostrovsky, J. (1971). The hippocampus as a spatial map: Preliminary evidence from unit activity in the freely-moving rat. *Brain Research, 34*(1), 171–175.

Pastalkova, E., Itskov, V., Amarasingham, A., & Buzsaki, G. (2008). Internally generated cell assembly sequences in the rat hippocampus. *Science, 321*(5894), 1322–1327.

Pillar, J., Malhotra, A., & Lavie, P. (2000). Post-traumatic stress disorder and sleep—what a nightmare! *Sleep Medicine, 4*(2), 183–200.

Robinaugh, D. J., LeBlanc, N. J., Vuletich, H. A., & McNally, R. J. (2014). Network analysis of persistent complex bereavement disorder in conjugally bereaved adults. *Journal of Abnormal Psychology, 123*(3), 510.

Sandell, C., Stumbrys, T., Paller, K. A., & Mallett, R. (2024). Intentionally awakening from sleep through lucid dreaming. *Current Psychology, 43*(21), 19236–19245.

Scarpelli, S., Bartolacci, C., D'Atri, A., Gorgoni, M., & De Gennaro, L. (2019). The functional role of dreaming in emotional processes. *Frontiers in Psychology, 10*, 459.

Schacter, D. L., Addis, D. R., & Buckner, R. L. (2008). Episodic simulation of future events: Concepts, data, and applications. *Annals of the New York Academy of Sciences, 1124*(1), 39–60.

Schacter, D. L., Addis, D. R., Hassabis, D., Martin, V. C., Spreng, R. N., & Szpunar, K. K. (2012). The future of memory: Remembering, imagining, and the brain. *Neuron, 76*(4), 677–694.

Schultz, W., Dayan, P., & Montague, P. R. (1997). A neural substrate of prediction and reward. *Science, 275*(5306), 1593–1599.

Seeber, M., Stangl, M., Vallejo Martelo, M., Topalovic, U., Hiller, S., Halpern, C. H., . . . & Suthana, N. (2025). Human neural dynamics of real-world and imagined navigation. *Nature Human Behavior,* https://doi.org/10.1038/s41562-025-02119-3.

Silva, D., Feng, T., & Foster, D. J. (2015). Trajectory events across hippocampal place cells require previous experience. *Nature Neuroscience, 18*(12), 1772–1779.

Slotnick, S. D., & Schacter, D. L. (2004). A sensory signature that distinguishes true from false memories. *Nature Neuroscience, 7*(6), 664–672.

Spanò, G., Pizzamiglio, G., McCormick, C., Clark, I. A., De Felice, S., Miller, T. D., . . . & Maguire, E. A. (2020). Dreaming with hippocampal damage. *eLife, 9,* e56211.

Spens, E., & Burgess, N. (2024). A generative model of memory construction and consolidation. *Nature Human Behaviour, 8*(3), 526–543.

Sunavsky, A., & Poppenk, J. (2020). Neuroimaging predictors of creativity in healthy adults. *NeuroImage, 206,* 116292.

Szpunar, K. K., Chan, J. C., & McDermott, K. B. (2009). Contextual processing in episodic future thought. *Cerebral Cortex, 19*(7), 1539–1548.

Szpunar, K. K., Watson, J. M., & McDermott, K. B. (2007). Neural substrates of envisioning the future. *Proceedings of the National Academy of Sciences, 104*(2), 642–647.

Voss, U., Holzmann, R., Tuin, I., & Hobson, A. J. (2009). Lucid dreaming: A state of consciousness with features of both waking and non-lucid dreaming. *Sleep, 32*(9), 1191–1200.

Wagner, U., Gais, S., Haider, H., Verleger, R., & Born, J. (2004). Sleep inspires insight. *Nature, 427*(6972), 352–355.

Watabe-Uchida, M., Eshel, N., & Uchida, N. (2017). Neural circuitry of reward prediction error. *Annual Review of Neuroscience, 40*(1), 373–394.

Winson, J. (1990). The meaning of dreams. *Scientific American, 263*(5), 86–97.

Wittmann, L., Schredl, M., & Kramer, M. (2006). Dreaming in posttraumatic stress disorder: A critical review of phenomenology, psychophysiology and treatment. *Psychotherapy and Psychosomatics, 76*(1), 25–39.

Yang, W., Sun, C., Huszár, R., Hainmueller, T., Kiselev, K., & Buzsáki, G. (2024). Selection of experience for memory by hippocampal sharp wave ripples. *Science, 383*(6690), 1478–1483.

Chapter 7. You Are What You Remember

Abrahao, K. P., Salinas, A. G., & Lovinger, D. M. (2017). Alcohol and the brain: Neuronal molecular targets, synapses, and circuits. *Neuron, 96*(6), 1223–1238.

Berntsen, D., Rubin, D. C., & Siegler, I. C. (2011). Two versions of life: Emotionally negative and positive life events have different roles in the organization of life story and identity. *Emotion, 11*(5), 1190–1201.

Boden, J. M., & Fergusson, D. M. (2011). Alcohol and depression. *Addiction, 106*(5), 906–914.

Calhoun, L. G., Tedeschi, R. G., Cann, A., & Hanks, E. A. (2010). Positive outcomes following bereavement: Paths to posttraumatic growth. *Psychologica Belgica, 50*(1–2), 125–143.

Çili, S., & Stopa, L. (2015). The retrieval of self-defining memories is associated with the activation of specific working selves. *Memory*, 23(2), 233–253.

de Snoo, M. L., Miller, A. M., Ramsaran, A. I., Josselyn, S. A., & Frankland, P. W. (2023). Exercise accelerates place cell representational drift. *Current Biology*, 33(3), R96–R97.

Driscoll, L. N., Duncker, L., & Harvey, C. D. (2022). Representational drift: Emerging theories for continual learning and experimental future directions. *Current Opinion in Neurobiology*, 76, 102609.

Ferrari, A., Gaggiotti, P., Silva, M., Veneroni, L., Magni, C., Signoroni, S., . . . & Massimino, M. (2017). Searching for happiness. *Journal of Clinical Oncology*, 35(19), 2209–2212.

Fredrickson, B. L. (1998). What good are positive emotions? *Review of General Psychology*, 2(3), 300–319.

Geva, N., Deitch, D., Rubin, A., & Ziv, Y. (2023). Time and experience differentially affect distinct aspects of hippocampal representational drift. *Neuron*, 111(15), 2357–2366.

Grella, S. L., Fortin, A. H., Ruesch, E., Bladon, J. H., Reynolds, L. F., Gross, A., . . . & Ramirez, S. (2022). Reactivating hippocampal-mediated memories during reconsolidation to disrupt fear. *Nature Communications*, 13(1), 4733.

Haubrich, J., Crestani, A. P., Cassini, L. F., Santana, F., Sierra, R. O., Alvares, L.D.O., & Quillfeldt, J. A. (2015). Reconsolidation allows fear memory to be updated to a less aversive level through the incorporation of appetitive information. *Neuropsychopharmacology*, 40(2), 315–326.

Hobson, J. A. (2009). REM sleep and dreaming: Towards a theory of protoconsciousness. *Nature Reviews Neuroscience*, 10(11), 803–813.

Keinath, A. T., Mosser, C. A., & Brandon, M. P. (2022). The representation of context in mouse hippocampus is preserved despite neural drift. *Nature Communications*, 13(1), 2415.

Kelly, J. F., & Greene, M. C. (2019). The reality of drinking and drug using dreams: A study of the prevalence, predictors, and decay with time in recovery in a national sample of U.S. adults. *Journal of Substance Abuse Treatment*, 96, 12–17.

Kersten, M., Swets, J. A., Cox, C. R., Kusumi, T., Nishihata, K., & Watanabe, T. (2020). Attenuating pain with the past: Nostalgia reduces physical pain. *Frontiers in Psychology*, 11, article 296.

Khatib, D., Ratzon, A., Sellevoll, M., Barak, O., Morris, G., & Derdikman, D. (2023). Active experience, not time, determines within-day representational drift in dorsal CA1. *Neuron*, 111(15), 2348–2356.

Klever, S. (2013). Reminiscence therapy: Finding meaning in memories. *Nursing*, 43(4), 36–37.

Koelsch, S. (2009). A neuroscientific perspective on music therapy. *Annals of the New York Academy of Sciences*, 1169(1), 374–384.

Lazar, A., Thompson, H., & Demiris, G. (2014). A systematic review of the use of technology for reminiscence therapy. *Health Education & Behavior*, 41(1Suppl), 51S–61S.

Lempert, K. M., Speer, M. E., Delgado, M. R., & Phelps, E. A. (2017). Positive autobiographical memory retrieval reduces temporal discounting. *Social Cognitive and Affective Neuroscience*, 12(10), 1584–1593.

Mankin, E. A., & Fried, I. (2020). Modulation of human memory by deep brain stimulation of the entorhinal-hippocampal circuitry. *Neuron*, 106(2), 218–235.

Mankin, E. A., Sparks, F. T., Slayyeh, B., Sutherland, R. J., Leutgeb, S., & Leutgeb, J. K. (2012). Neuronal code for extended time in the hippocampus. *Proceedings of the National Academy of Sciences, 109*(47), 19462–19467.

Mau, W., Hasselmo, M. E., & Cai, D. J. (2020). The brain in motion: How ensemble fluidity drives memory-updating and flexibility. *eLife, 9*, e63550.

McAdams, D. P., & McLean, K. C. (2013). Narrative identity. *Current Directions in Psychological Science, 22*(3), 233–238.

McAdams, D. P., Reynolds, J., Lewis, M., Patten, A. H., & Bowman, P. J. (2001). When bad things turn good and good things turn bad: Sequences of redemption and contamination in life narrative and their relation to psychosocial adaptation in midlife adults and in students. *Personality and Social Psychology Bulletin, 27*(4), 474–485.

Miller, A. P., Baranger, D. A., Paul, S. E., Garavan, H., Mackey, S., Tapert, S. F., . . . & Bogdan, R. (2024). Neuroanatomical variability and substance use initiation in late childhood and early adolescence. *JAMA Network Open, 7*(12), e2452027.

Muir, J., Lin, S., Aarrestad, I. K., Daniels, H. R., Ma, J., Tian, L., . . . & Kim, C. K. (2024). Isolation of psychedelic-responsive neurons underlying anxiolytic behavioral states. *Science, 386*(6723), 802–810.

Munezawa, T., Kaneita, Y., Osaki, Y., Kanda, H., Ohtsu, T., Suzuki, H., . . . & Ohida, T. (2011). Nightmare and sleep paralysis among Japanese adolescents: A nationwide representative survey. *Sleep Medicine, 12*(1), 56–64.

Nandrino, J. L., Gandolphe, M. C., & El Haj, M. (2017). Autobiographical memory compromise in individuals with alcohol use disorders: Towards implications for psychotherapy research. *Drug and Alcohol Dependence, 179*, 61–70.

Noetel, M., Sanders, T., Gallardo-Gómez, D., Taylor, P., del Pozo Cruz, B., Van Den Hoek, D., . . . & Lonsdale, C. (2024). Effect of exercise for depression: Systematic review and network meta-analysis of randomised controlled trials. *BMJ, 384*, e075847.

Pavlova, M., Lund, T., Nania, C., Kennedy, M., & Graham, S. (2022). A randomized controlled trial of a parent-led memory-reframing intervention. *Journal of Pain, 23*(2), 263–275.

Rechavi, Y., Rubin, A., Yizhar, O., & Ziv, Y. (2022). Exercise increases information content and affects long-term stability of hippocampal place codes. *Cell Reports, 41*(8).

Rey, H. G., Gori, B., Chaure, F. J., Collavini, S., Blenkmann, A. O., Seoane, P., . . . & Quiroga, R. Q. (2020). Single neuron coding of identity in the human hippocampal formation. *Current Biology, 30*(6), 1152–1159.

Rimmele, U., Besedovsky, L., Lange, T., & Born, J. (2015). Emotional memory can be persistently weakened by suppressing cortisol during retrieval. *Neurobiology of Learning and Memory, 119*, 102–107.

Sacharin, V. (2009). The influence of emotions on cognitive flexibility. PhD diss., University of Michigan.

Sadeh, S., & Clopath, C. (2022). Contribution of behavioural variability to representational drift. *eLife, 11*, e77907.

Samide, R., & Ritchey, M. (2021). Reframing the past: Role of memory processes in emotion regulation. *Cognitive Therapy and Research, 45*(5), 848–857.

"Science: The fleeting flesh." *Time*, October 11, 1954, https://time.com/archive/6869550/science-the-fleeting-flesh.

Stieger, S., & Kuhlmann, T. (2018). Validating psychometric questionnaires using experience-sampling data: The case of nightmare distress. *Frontiers in Neuroscience, 12*, article 901, https://doi.org/10.3389/fnins.2018.00901.

Tedeschi, R. G., & Calhoun, L. G. (1996). The Posttraumatic Growth Inventory: Measuring the positive legacy of trauma. *Journal of Traumatic Stress, 9*, 455–471.

Trouche, S., Sasaki, J. M., Tu, T., & Reijmers, L. G. (2013). Fear extinction causes target-specific remodeling of perisomatic inhibitory synapses. *Neuron, 80*(4), 1054–1065.

Williams, S. E., Ford, J. H., & Kensinger, E. A. (2022). The power of negative and positive episodic memories. *Cognitive, Affective, & Behavioral Neuroscience, 22*(5), 869–903.

Wolf, T., Strack, V., & Bluck, S. (2021). Adaptive and harmful autobiographical remembering after the loss of a loved one. *Aging and Mental Health, 25*(3), 460–469.

Woods, B., O'Philbin, L., Farrell, E. M., Spector, A. E., & Orrell, M. (2018). Reminiscence therapy for dementia. *Cochrane Database of Systematic Reviews, 3*, article CD001120.

Yong, E. (2021). Neuroscientists have discovered a phenomenon that they can't explain. *The Atlantic*, June 9.

Zaki, Y., Pennington, Z. T., Morales-Rodriguez, D., Bacon, M. E., Ko, B., Francisco, T. R., ... & Cai, D. J. (2025). Offline ensemble co-reactivation links memories across days. *Nature, 637*, 145–155.

Ziv, Y., Burns, L. D., Cocker, E. D., Hamel, E. O., Ghosh, K. K., Kitch, L. J., ... & Schnitzer, M. J. (2013). Long-term dynamics of CA1 hippocampal place codes. *Nature Neuroscience, 16*(3), 264–266.

Chapter 8. So Long Lives This, and This Gives Life to Memory

Buzsáki, G., & Moser, E. I. (2013). Memory, navigation and theta rhythm in the hippocampal-entorhinal system. *Nature Neuroscience, 16*(2), 130–138.

Carrillo-Reid, L., Han, S., Yang, W., Akrouh, A., & Yuste, R. (2019). Controlling visually guided behavior by holographic recalling of cortical ensembles. *Cell, 178*(2), 447–457.

Chen, B. K., Murawski, N. J., Cincotta, C., McKissick, O., Finkelstein, A., Hamidi, A. B., ... & Merfeld, E. (2019). Artificially enhancing and suppressing hippocampus-mediated memories. *Current Biology, 29*(11), 1885–1894.

Coffman, B. A., Clark, V. P., & Parasuraman, R. (2014). Battery powered thought: Enhancement of attention, learning, and memory in healthy adults using transcranial direct current stimulation. *NeuroImage, 85*, 895–908.

Creed, M., Pascoli, V. J., & Lüscher, C. (2015). Refining deep brain stimulation to emulate optogenetic treatment of synaptic pathology. *Science, 347*(6222), 659–664.

Deiana, S., Platt, B., & Riedel, G. (2011). The cholinergic system and spatial learning. *Behavioral Brain Research, 221*, 389–411.

Demas, J., Manley, J., Tejera, F., Kim, H., Barber, K., Martínez Traub, F., ... & Vaziri, A. (2021). Volumetric calcium imaging of 1 million neurons across cortical regions at cellular resolution using light beads microscopy. *bioRxiv*, 2021.02.21.432164.

Dennett, D. (2002). Dangerous memes. TED conference, February, https://www.ted.com/talks/dan_dennett_dangerous_memes.

Eich, T. S., & Metcalfe, J. (2009). Effects of the stress of marathon running on implicit and explicit memory. *Psychonomic Bulletin & Review, 16*, 475–479.

Eichenbaum, H. (2016). Still searching for the engram. *Learning & Behavior, 44*(3), 209–222.

Erickson, K. I., Voss, M. W., Prakash, R. S., Basak, C., Szabo, A., Chaddock, L., . . . & Kramer, A. F. (2011). Exercise training increases size of hippocampus and improves memory. *Proceedings of the National Academy of Sciences, 108*(7), 3017–3022.

Foster, J. K., Lidder, P. G., & Sünram, S. I. (1998). Glucose and memory: Fractionation of enhancement effects? *Psychopharmacology, 137*, 259–270.

Gold, P. E., Vogt, J., & Hall, J. L. (1986). Glucose effects on memory: Behavioral and pharmacological characteristics. *Behavioral Neural Biology, 46*, 145–155.

Hasselmo, M. E. (2015). If I had a million neurons: Potential tests of cortico-hippocampal theories. *Progress in Brain Research, 219*, 1–19.

Josselyn, S. A., & Tonegawa, S. (2020). Memory engrams: Recalling the past and imagining the future. *Science, 367*(6473).

Lapp, J. E. (1981). Effects of glycemic alterations and noun imagery on the learning of paired associates. *Journal of Learning Disabilities, 14*, 35–38.

Linssen, A.M.W., Vuurman, E.F.P.M., Sambeth, A., & Riedel, W. J. (2011). Methylphenidate produces selective enhancement of declarative memory consolidation in healthy volunteers. *Psychopharmacology, 221*, 611–619.

Lüscher, C., Pascoli, V., & Creed, M. (2015). Optogenetic dissection of neural circuitry: From synaptic causalities to blueprints for novel treatments of behavioral diseases. *Current Opinion in Neurobiology, 35*, 95–100.

Pastuzyn, E. D., Day, C. E., Kearns, R. B., Kyrke-Smith, M., Taibi, A. V., McCormick, J., . . . & Yoder, N. (2018). The neuronal gene *Arc* encodes a repurposed retrotransposon gag protein that mediates intercellular RNA transfer. *Cell, 172*(1–2), 275–288.

Phelps, E. A., & Hofmann, S. G. (2019). Memory editing from science fiction to clinical practice. *Nature, 572*(7767), 43–50.

Poo, M. M., Pignatelli, M., Ryan, T. J., Tonegawa, S., Bonhoeffer, T., Martin, K. C., . . . & Rudenko, A. (2016). What is memory? The present state of the engram. *BMC Biology, 14*, 1–18.

Robinson, N. T. M., Descamps, L. A., Russell, L. E., Buchholz, M. O., Bicknell, B. A., Antonov, G. K., . . . & Häusser, M. (2020). Targeted activation of hippocampal place cells drives memory-guided spatial behavior. *Cell, 183*(6), 1586–1599.

Roig, M., Nordbrandt, S., Geertsen, S. S., & Nielsen, J. B. (2013). The effects of cardiovascular exercise on human memory: A review with meta-analysis. *Neuroscience & Biobehavioral Reviews, 37*(8), 1645–1666.

Smith, M. A., Riby, L. M., Eekelen, J.A.M.V., & Foster, J. K. (1998). Glucose enhancement of human memory: A comprehensive research review of the glucose memory facilitation effect. *Neuroscience & Biobehavioral Reviews, 22*, 335–345.

Stern, S. A., & Alberini, C. M. (2013). Mechanisms of memory enhancement. *Wiley Interdisciplinary Reviews: Systems Biology and Medicine, 5*(1), 37–53.

Suthana, N., & Fried, I. (2014). Deep brain stimulation for enhancement of learning and memory. *NeuroImage, 85,* 996–1002.

Suthana, N., Haneef, Z., Stern, J., Mukamel, R., Behnke, E., Knowlton, B., & Fried, I. (2012). Memory enhancement and deep-brain stimulation of the entorhinal area. *New England Journal of Medicine, 366*(6), 502–510.

Sweis, B. M., Mau, W., Rabinowitz, S., & Cai, D. J. (2021). Dynamic and heterogeneous neural ensembles contribute to a memory engram. *Current Opinion in Neurobiology, 67,* 199–206.

INDEX

AA (Alcoholics Anonymous), 171
academia, criticisms in, 12–13
activity-dependent c-Fos upregulation, 25
addiction, 80, 164, 177; binge drinking,
   168–69; changing understanding of
   experience, 173; connection as opposite
   of, 170; hangover-induced anxiety,
   167–68; intervention by friends, 170;
   memory of bender, 167; regret of not
   seeking help, 169–70; sobriety and, 174,
   175
adeno-associated viruses, 25
alcohol: binge drinking, 167–69; drowning
   memories with, 170; experience with,
   163; lucid dreaming in recovery, 207n2;
   sleep and, 163; sobriety and, 174, 175
alcohol use disorder, family struggle with,
   171
Alzheimer's disease, 11, 93
amnesia: drugs blocking memory
   formation, 185–86; hippocampus, 137;
   memory and, 92–93; Ryan on bringing
   memory back from, 90–91
amnesic patient (E.P.), bilateral damage to
   hippocampus, 61, 62
"Amnestic Trace, The," 88
amygdala: brain's emotional loci, 73,
   97; engram location, 187; memory
   formation, 97
anthropomorphic conjecture, mouse
   recalling memory as, 7–8
anxiety, 6, 114, 116, 164, 189; animal behavior,
   123; disorders, 81; grief and, 176; panic

attack and, 4; reactivating positive
   memories, 124
Aristotle, on dreams, 136
atoms, change in human body, 160
auto-mechanics, memory researchers as, of
   brain, 3
Axel, Richard, on artificially activating cells,
   109

Barrymore, Drew, character in *50 First
   Dates*, 61
Begley, Sharon, on something incredible
   waiting to be known, 183
Bell, Dugald, on absence of evidence,
   204n5
Benchenane, Karim, memory formation in
   sleeping animal, 206n2
bidirectional switch, positive and negative
   engrams, 108
BIOCOM 200 computerized image-
   processing, 60
biology, world of unrest, 156
Blade Runner 2049 (movie), 96–97
Bontempi, Bruno, on reorganization of
   brain circuitry, 60
Born, Jan, on regulating emotional
   responses of negative memory, 171
Boston Marathon, 64; bombing at 2013,
   76–77, 85, 93; dream of running, 76;
   memory of running, 77–79, 85–86,
   93; unofficial runners as bandits,
   203n2
Boston University, 14, 28

brain: activating cells and jump-starting recollection, 87–89; changing moment to engram, 156–57; domino effect of external stimuli to memory, 3–4; finding inconsistency, 142; fly, 5; hearing the music of the, 192; human, 5; information for flexible future use, 99–100; jump-starting cascade of events, 187; Lipton on study of, 28; memories as building blocks, 136–37; memory location in, 191; modifying memories by time and experience, 162; negative valence systems of, 121; neuroscience and, 115; neuroscience of broken, 114–15; predicting loss of individual, 148; preplay and replay of memory, 138–39; promoting growth of new cells, 124; prospective coding, 139; replaying moments during sleep, 139–40

Brain and Cognitive Sciences building, 1; MIT's graduate program, 21; Ramirez' graduate school acceptance, 30

brain cells, Deisseroth Lab, 45–46

brain circuitry, paper on reorganization of, 60

brain stimulation, memory editing, 184–86

Buckner, Randy, on mental time-travel, 152

Cajal, Santiago Ramón y: Golgi and, 47; human as sculptor of own brain, 182; on improving Golgi stain method, 39; landscape of brain, 38; on memory, 43

Capital Grille, experience at, 158–59, 162

cells: activating, and jump-starting recollection, 87–88; response to name, 160

change: biology of, 157, 176–77; concept, 156; drift, 161–62; exercise remodeling memory circuits of brain, 173; narrative identity, 165–66; negative memories, 171–72; physics, 160; posttraumatic growth, 174; power of memory, 178–79; representational drift, 161

Channelrhodopsin-2 (ChR2), 11; explaining in Spanish, 55; studies of, 45, 46, 53; test tube label, 21

channels, ions and cells, 46

Chen, Briana, on reactivating positive memories, 124

childhood memories, 56–57

chloride ions, channels, 46

circuits, teams of cells forming, 58

cognitive behavioral therapy (CBT): major depressive disorder, 122; remodeling memory circuits of brain, 173; treating disorders, 124

cognitive psychology, knowledge of someone's death, 147

collaboration, 36–37

collective memories, 79

Columbia University, 42, 74, 109

Corkin, Suzanne, study of HM, 42

Cowansage, Kiriana, on memories in numerous brain areas, 97

creativity, imagination, 141–42

Creed, Meaghan, optogenetics abolishing pathological behavior in rodents, 193

criticisms, academia and, 12–13

Curie, Marie, radioactivity, 51

Dark Knight Rises, The (movie), 57

death, 164; Xu Liu, 143–46, 165

deep brain stimulation (DBS): memory editing, 185; optogenetic-inspired procedures, 193

Deese-Roediger-McDermott paradigm, false memory generation, 102

Deisseroth, Karl: on optogenetics, 45; on use of optogenetics, 62–63, 206n5

de Laviléon, Gaeton, memory formation in sleeping animal, 206n2

Delgado, Mauricio, on active recall of positive memories, 124–25

dementia, 104

Denny, Christine, on neural mechanisms of learning and memory, 74

depression, 11, 116; animal behavior, 123; grieving the loss of future, 148–49;

for, 118–19; Research Domain Criteria (RDoC) project classifying, 120–22

mental health disorders, redefining, 116–17

Mighty Morphin Power Rangers action figure, 57

Miller, Glenn, "In the Mood," 13

Milner, Brenda, study of HM, 42

misinformation effect, Loftus on, 100–101

MIT (Massachusetts Institute of Technology), 1, 7, 14, 27; Great Dome, 29

Molaison, Henry (HM): bike accident and epileptic convulsions of, 41–42; bilateral damage to hippocampus, 62; experience of, 58; memories of, 59; memories of childhood, 59; study by Milner and Corkin, 42

Monfils, Marie-H, on reconsolidation window of memory, 84

motor memory, 59

mouse: experimentation, 1–2; forming "neutral" memory, 106; hippocampus, 2; reactivating positive memories, 190; recall of false memory, 106–7; stimulating hippocampus, 123–24. *See also* rodents

music: as backdrop of memory, 65–67; rule in lab, 54

Nader, Karim, on neural mechanisms underpinning reconsolidation, 83

narrative identity, process of, 165–66

National Institute of Mental Health (NIMH), 119, 120

National Institutes of Health (NIH), 164, 196; Early Independence Award, 180, 188; funding for Ramirez Lab, 196–97

Natronomonas pharaonic, NpHR (halorhodopsin), 63

Nature (journal), 11, 60, 61

negative memory: changing, 171–72; positive memories and, 172–73

negative valence systems, brain, 121, 205n3

neural activity, memories of people, 147–48

neurons, drugs and, 44

neuroscience: brief history of, 39–40; broken brains and broken thoughts, 114–15; court of law and, 195; growing field, 51; manipulating memory, 18; memories as dynamic reconstructions of past, 96; memory and amnesia, 92–93; memory manipulation, 75, 116; suppression of old memories, 82; twenty-first-century, 17

neuroscientists, playing hide-and-seek with memories, 91–92

New York University, 83, 84, 139

1984 (Orwell), 137

Nobel Prize, 38

Northwestern University, 126, 128, 145

NpHR (halorhodopsin), *Natronomonas pharaonic*, 63

Oak Ridge National Laboratory, 160

Ogawa, Seiji, scientific idea of fMRI, 204–5n2

"on/off" switch: goal of Ramirez Lab, 189; memory, 74

"on" switch, memory, 74

optic fibers (OF), labels on rodents, 52–54

optogenetics, 44; depression-related symptoms in mice using, 206n5; development of, 45; Goshen and team on use, 62–63; holographic approach, 110; imaging and, 48; listening in on animal brain cells, 189–90; meaning of, 45; turning cells on and off, 47

Orwell, George, *1984*, 137

Owen, Adrian: responsive brains of patients, 36; tests measuring vegetative state, 35–36

Padilla-Coreano, Nancy, on memory extinction, 81

panic attack, 4, 114

parents: background of Ramirez', 30–31, 30–33; Ramirez sharing graduate school acceptance with, 30; support of, 31–33